ETERNAL LIFE - WHY YOU SHOULD EXPECT TO LIVE FOREVER

FINDING UNEXPECTED STRENGTH WHEN THE WORLD TRAMPLES ON YOUR FAITH

JOHN ZACHARY

Harvard House
PUBLISHING

ETERNAL LIFE

WHY YOU SHOULD EXPECT TO LIVE FOREVER!

Finding unexpected strength when the world tramples on your faith!

John Zachary

Copyright © 2020 by Harvard House

Eternal Life – Why you should expect to live forever

All rights reserved. No part of this publication may be reproduced, distributed or transmitted in any form or by any means, including photocopying, recording, or other electronic or mechanical methods, without the prior written permission of the publisher, except in the case of brief quotations embodied in critical reviews and certain other noncommercial uses permitted by copyright law.

Bible quotations are from the New American Standard Bible (NASB) unless otherwise specified when quoted.

Cover Design by www.delaney-designs.com
All Rights Reserved
ISBN: 978-1-951885-04-5

I dedicate this book series to Pastor Charlie, who found 14,000 days hidden around the lives of Moses and Joshua. Many people have read these texts across thousands of years. However, no one knew of the 14,000 days! This hidden period comes from God leading Israel. As a result, this evidence supports awe-inspiring guidance for Moses and Joshua. For sure, they did not know of this!

I also dedicate this book to my parents. They added to book one in the Expect to Live Forever book series, <u>Secrets - never heard until now - of the Book of Revelation</u>, chapter four.

CONTENTS

Suggested Ways to Read this Book	ix
Preface	xi
1. Can we Know that God Exists?	1
2. The Ultimate Premise and Question	7
3. What existed before the Universe Began?	14
4. Atheism: Its Quest for an Eternal Natural Universe	16
5. Pantheism:	32
6. Monotheism:	40
7. Using Logic to Probe	48
8. When Was Daniel Written?	51
9. What events were foretold?	56
10. Upon what day of the week did each event occur?	59
11. Repeatability of the Biblical Constants	61
12. Insights from the Mystical 14,000 Days	69
13. Jewish Temples: Purpose and Controlled Destruction	74
14. Burning Down Solomon's Temple	80
15. Scientific Date for the Beginning of the Prophecy	87
16. Biblical Date for the End of the Sixty-Ninth Week	92
17. Scientific Date for Burning Down Herod's Temple	97
18. Evidence that Validates AD 32	102
19. Jesus' Second Coming Promise	123
20. Do the Time-Based Texts Foretell Future Events?	135
21. Global-wide Christian Revival	143
Please Review My Book	145
Acknowledgment	147
Expect to Live Forever Book Series	149
About the Author	151
Bibliography	153
Notes	157

SUGGESTED WAYS TO READ THIS BOOK

Book two – Expect to Live Forever book series

Chapters one through six investigate various worldviews with known science. These chapters increase your knowledge of other worldviews in light of the Christian worldview.

- For example, is the Atheist worldview valid, based on known science? Does Atheism have flaws?
- Can you defend your faith against non-Christian worldviews?
- Will your worldview find support with known science?

Perhaps you only want to learn how the Bible foretells of specific future events that will occur on an exact day, centuries into the future? Then I suggest you begin in chapter six. When you complete the reading of chapter twenty-one, you will realize the value of the first five chapters.

For sure, we cannot verify worldviews with scientific experiments. In contrast, the evidence of prophetic texts that predict specific events will occur on an exact day, centuries into the future, do not require scientific experiments. Why?

For certain, no created entity knows the future. In contrast, only a transcendent Being can foretell the future and line up the final 14,000th day to an event with high-level spiritual meaning. If you take the time to study and learn these phenomena, you will begin to realize that you should expect to live forever.

If you want to know why I have confidence in this information, I suggest that you read book one in the Expect to Live Forever book series, *Secrets - never heard until now - of the Book of Revelation*.

PREFACE

If you knew, without question, that you are going to live forever, would that change how you live your life?

We know that no one can predict the exact day of a future event, several hundred years before the foretold incidents occur. Also, the texts must be recorded on paper, centuries earlier. However, I have found it to be true. Therefore it is fitting to ask, *"Who is the Source of those texts?"*

For sure, the Source knows the details of those future events. In fact, the Source must transcend time. Time, of itself, would be meaningless to the Source.

Some people may inquire, *"Is this evidence of a time traveler?"*

Three things answer this question.

First, Einstein's special relativity and general relativity strongly suggest humans cannot time travel, ever. Despite this knowl-

Preface

edge, some people will continue to believe that time travel will become possible.

Second, in contrast with people who would continue to believe in time travel, we have more evidence that the texts are not from a time traveler. This evidence comes from the details of the foretold events. Why?

Consider that the predicted events are strictly related to spiritual knowledge. Therefore, we logically conclude the Source is a Spiritual Being of the highest order. To this Source, time and space are meaningless. For sure, it would be easy for the source to predict specific future events that will occur on an exact date, many centuries into the future.

Finally, the third fact comes from the details written into book one of the Expect to Live Forever book series. If you want those details, please read <u>Secrets – never heard until now – of the Book of Revelation</u>.

Perhaps you are wondering how we would know that the texts were recorded on paper, centuries before each event. The Dead Sea Scrolls validate the documents pre-existed the foretold events. You need one more piece of information. The recorded words reveal the source is God. The texts claim that a prophet received them from an angel named Gabriel.

We can begin to grasp a fantastic idea. Foretelling exact events into the future means that God wants us to know that He exists. In truth, God is revealing Himself to us, which leads to the conclusion: "You should expect to live forever!"

In this book, you gain insight into the false viewpoints offered by Atheism. Among these false claims, atheists claim that science validates their views. However, science refutes the atheist perspective, provided you want to know.

Please realize, God only reveals Himself so you can know that you are exceptional. You are so excellent that you should expect to live forever. You can have confidence in this conclusion.

1
CAN WE KNOW THAT GOD EXISTS?

WE LIVE IN A TIME WHEN ALL WE HEAR IS THAT RELIGION AND science do not get along. At the extreme level, so-called scientific atheists claim that God does not exist!

Consider this fundamental question.

Isn't there any science-based data to prove that God is dead?

The correct answer is no. If that data were available, every scientist would be an atheist!

Perhaps you are thinking, *"Just show me an example of a world-class scientist that is Christian?"*

Consider Dr. John Polkinghorne, who is a theoretical physicist and Anglican priest. Why would a world-class scientist become an Anglican priest?

Before deciding to become an Anglican priest, Dr. Polkinghorne played a role in finding the quark[1]. He has authored five books on physics as well as another twenty-six books on the relationship between science and faith. You may be surprised to know that he has a strong belief in Jesus rising from the dead!

Why would a world-class scientist believe in God and miracles?

Brilliant people will continue to believe in God. As you read into this book, you will learn that Atheism logically leads to a supernatural realm. Why?

To validate your worldview, you must show how the source of the universe is eternal. However, Atheism only accepts the natural realm. The critical issue is that the natural world must be endless, but "known science" reveals that every natural thing will not last forever. Atheism cannot solve this problem. Atheism relies on unbelievable amounts of blind faith in the impossible!

In contrast, the supernatural realm will never stop working. In this book, we will see how known science and logic unveil the spiritual realm. In simple language, the source must be greater than our universe to explain the world and the fact that the cosmos is dying. I will give more details on this in later chapters.

This chapter introduces you to three methods that unveil the spiritual world. Before I write about that, it is crucial to grasp that we must have data of some type. Because data lets us know that the spiritual world is a fact, just like the air we breathe!

Some people may not like using the word "fact" for the spiritual realm. Another way of saying this would be that the data suggests the spiritual realm is factual. We study the data to find out if we can have confidence in our conclusions. If we have confidence in the spiritual realm, then we must also answer the question. *"What is faith?"*

Method One: Texts that Foretell the Future

Do you think it is possible to write words on paper that will predict the future? Not the events that will take place in the next hour or the next minute, but centuries later. For example, *"What will happen in three hundred years? Also, what will be the military action on that day?"*

Of course, this is not possible for beings such as ourselves. Or aliens in spaceships, if they exist! Also, it is not viable for unseen spiritual beings such as angels or demons! For sure, not even the Archangel Gabriel would know the future as some may think! Because time and space do not let created beings know the future. Creatures, in contrast with the Creator, do not have access to events in the distant future. Time controls everything that created entities do.

Only the Creator can foretell the future. In this book, we find the details that strongly support, even verify, that time-based texts predicted specific events that happened on a precise date, centuries into the future.

Moreover, we discover that foretold events have unique *high-level spiritual meanings*. These events teach us about the spiritual realm and require the event to occur on an exact date, centuries into the future. A summary would mean at least four things about this transcendent Being.

- This Being transcends time!
- This Being exists outside of our universe, where time and space mean nothing.
- This Being only foretells of events with high-level spiritual meaning!

> NOTE: In general, all other events do not have God-like control. You have free will to live as you please.

- To cause the specific event foretold to happen on an exact date, centuries into the future, this Being will control humans to bring about that prophesied event on an exact day. I must emphasize that this Being only controls events with high-level spiritual meaning! The dates of these foretold events strongly support this viewpoint. This evidence unveils the spiritual realm, like the air we breathe!

As a result, we learn priceless information about ourselves. We can logically conclude, *"We can expect to live forever!"*

What else can we find from words that foretell the distant future?

This method also lets us analyze differing beliefs. Are the differing "beliefs" true or false?

This approach exposes the shortcomings of what so-called scientific atheists believe[2].

Method Two: Observation of the Effects of Spiritual Beings

Most people would wonder if it is possible to view events "caused by" or "influenced by" a spiritual being. The notion, of course, seems hard to believe, but it is possible. How?

Because I was one of five people who saw such an event, which I refer to as the *Psychic Bird Story*! This story is recorded in book four in the Expect to Live Forever book series, <u>Secrets - *never heard until now* - of the Spiritual Realm</u>. The five of us viewed the effects of a spiritual being during a Bible study. The demonic spirit made a parakeet in a cage squawk two to three times each second. Also, the bird beat its wings as fast as it could. This spiritually influenced event took about five minutes. The purpose was to distract from my teaching; "spiritual beings possess genuine psychics!"

Based on that, we can conclude that a spiritual realm exists. The evidence is sound. We can calculate our confidence level for the spiritual realm. There is only one chance out of 10,000 that the bird decided to do this while I taught about demons. The calculation comes from the observation of five people. We all saw it and heard it happen. I observed people being distracted by the bird! I suggest you read the *Psychic Bird Story* recorded in book four of the Expect to Live Forever book series.

Consider this question: *"Are there other stories where a group of two or more people observed the effects of a spiritual being?"*

If you read all the books in this Expect to Live Forever book series, you will be able to grasp more about the spiritual realm. Hence, you need to prepare yourself so you do not get deceived by religious deceivers or by Atheists.

I am sure there are other meetings where people have seen the effects of spiritual beings. I would expect to find more evidence!

Finally, this method is validated by scientific dating specific foretold events to occur on an exact date, centuries into the future. Therefore, two examples are included in the Expect to Live Forever book series.

Method Three: Scientific Principles and Data Reveal Spiritual Truth

Will the universe last forever?

It is a fact that the sun will eventually stop producing light and heat. Why?

Our sun has limited amounts of atomic energy. The sun, at the end of its life, will become colder than ice. Many stars have already used up their nuclear power. Then they cool off over eons of time, which will happen to every heavenly light in the cosmos.

Scientists[3] came to this conclusion in the 1850s. Since then, science has continued to prove this claim. Every scientist knows that the universe is doomed. It has an expiration date. NASA scientist Michael Pauken[4] writes:

> *"I hope you don't get depressed when I tell you that the universe is going to end one day because of the 'increase in entropy' principle."*

This NASA scientist validates that the universe cannot take care of itself. People agree with Michael Pauken's statement due to the vast amounts of data. Therefore, it makes sense to ask:

> *"What does this science-based information reveal about finding spiritual truth?"*

2

THE ULTIMATE PREMISE AND QUESTION

Since the universe cannot keep itself going forever, we must ask. *"Is it possible that our universe would be able to create another world?"*

The logical answer is no. From the start of this universe, it has been wearing out like old clothes. Since it only wears out, how can it create something else?

Consider this question *"Can I prevent myself from dying?"* Since everyone dies, we know the answer can only be no.

Based on the above conclusions, what would be required to create our universe?

The source of our universe is going to be very powerful in creating the world in which we live. Perhaps the most crucial aspect of the source is that it must be eternal. We discover the critical point that eternity going back forever confronts the Atheist view of the universe. So that you understand, the top-level atheist thinkers are asking you to believe in the impossible.

Consider this critical question that the Atheist worldview cannot answer based on known science. *"How did the universe begin?"*

If you take the time to honestly think through the Atheist view of how the universe began, you realize there is only one possible conclusion. The Atheist thinkers want you to *"believe in a natural universe that is like God!"* Let me emphasize that the source of the universe must be eternal.

There is no way of getting around the idea of the source being eternal, which becomes evident when you read the next three chapters. At this time, we know that the source would have to exist forever, without beginning or end. A natural universe is not going to last forever.

Remember the NASA scientist I quoted? He verified that the universe is dying. He referred to the "principle of increasing entropy." This is the science-based reason that causes our world to die. This principle is the law of the universe. For sure, everything in our universe will die.

Since this principle controls this universe, would it control the source of our world?

Based on logic, the principle of increasing entropy[1] will not control the source of our universe. We can reason that the Creator of the cosmos must exist forever! This Source exists above natural law!

Think about the saying, "above natural law." What does that mean?

Consider that the phrase, above natural law, is the same as using the word, supernatural. Logic points us toward a God-like source. There is no other answer.

Based on the principle of increasing entropy, we logically get to the following premise.

> ### The Premise
>
> **The natural universe cannot exist without an eternal source!**

Perhaps I need to remind you that the foundation comes from proven scientific principles[2]. Consider a better way of writing this idea.

> ### The Premise
>
> **The natural universe cannot exist without a Supernatural source!**

The Source will have the power to create our universe, and this Source must be eternal. Logic, of itself, leads us to ask two logical questions.

> ### Logical Questions
> **Will we see miracles take place in our universe?**
> **Will we see events that spiritual beings caused?**

For these questions, the logical answer is YES!

Using the Premise to Evaluate Belief Systems

WHAT DO THE PREMISE AND THE LOGICAL QUESTIONS REVEAL TO US?

To begin, we perceive the idea of a supernatural realm as a logical idea based on known science. Plainly stated, known science points us to God's existence as a logical conclusion.

Another issue immediately arises since we are thinking about the spiritual realm. How can we know who or what God is?

To arrive at a credible conclusion, we must use evidence and logic. Also, we must define the different views of God that exist on the Earth. Is that even possible?

Since the universe requires a supernatural Source that is eternal, it would be ideal to find out how different religions view the cosmos. We will discover how religions view the universe, of itself, to be the perfect way of finding what is true of the supernatural realm.

The Universe as Defined by Major Belief Systems

How does each religion view the universe?

We learn that this question divides all the religions on Earth into three categories. To do this, we ask two simple questions.

"Is God "IN" the Universe?" or "Is God "OUTSIDE" the universe?" A third viewpoint that claims to have an answer to these questions comes from the Atheist worldview. They do not believe that God exists.

Let us begin with God existing "IN" the Universe.

Is God the Universe?

We classify religions that believe God is the Universe as Pantheism. In general, people with this viewpoint do NOT talk about heaven and hell. Instead, they talk about dying, then coming back to earth as someone else. Perhaps you know of the word reincarnation. Plainly stated, after you die, you return to the Earth as another person. You are attempting to become a better person, or perhaps a person of the highest moral and ethical character.

Perhaps you know of the movie, Groundhog Day? That movie represents this viewpoint. Actor Bill Murray wakes up every day on groundhog day (February 2) for eternity until he becomes perfect. It is one of my favorite movies as it makes me laugh a lot.

Does Pantheism have any flaws in its view of the *Universe*?

You probably noticed that I capitalized, *Universe*. Why?

Pantheist beliefs view the physical *Universe* as God. Pantheism requires the Universe to be eternal. In contrast, known science in the previous chapter reveals the universe is dying. How will Pantheism find a way around this problem? I will entertain this question when writing about Pantheist beliefs.

Is God Outside the Universe?

Now think about religions that believe God exists outside the universe. We would say, "God transcends the universe." We call this set of beliefs, "Transcendent."

In the previous chapter, we learned that the universe is dying. How will transcendent religions deal with this problem?

We instantly find this is not a problem because God creates the universe. This belief system does not have a problem with known science and the end of the cosmos. The fact of the universe dying aligns with the teachings of all these religions.

People with transcendent beliefs talk about heaven and hell. In general, you do not get a second chance offered by the belief in reincarnation.

What is the third belief about God?

Atheism is a belief about ultimate reality. Perhaps you are wondering what is meant by the phrase, "Ultimate reality?"

Ultimate reality is about what happens to you when you die. If you have a soul, you continue to live after death. What if all beliefs are fairy tales?

To Atheists, when you die, your life is completely over. Are they right? Does Atheism have any flaws in its view of the universe?

The universe, how it began, is the brick wall for Atheism. Imagine running as fast as you can into a solid brick wall that is built with reinforced steel and is ten meters thick. The only thing that stops would be your bruised and bloody face and body. This mental picture displays the reality of the Atheist worldview.

In the previous chapter, we learned that known science reveals the universe is dying. How will Atheism find a way around this problem? I will entertain this question when writing about Atheist beliefs.

In my experience of discussing this with Atheists, they choose to ignore this problem. However, I will delve deeper into this in the chapter on Atheism.

Briefly, the existence of our universe requires the Source of the cosmos to be eternal. For Atheism, there is no spiritual realm. This means that a natural world must be endless! Perhaps better stated, Atheists are required to believe there is a natural cosmos that is supernatural.

Now a severe question because we know our universe is dying. How will we find a natural world that is eternal?

Our universe began many billions of years ago based on what NASA people teach with their data. When this viewpoint is analyzed, I will include the position of Richard Dawkins. He wrote *The God Delusion*. He is a firm believer in the idea that evolution explains everything.

In this book, I require science-based rules. Without data, we are in the dark. We consider each viewpoint but only with data.

Perhaps you think that beliefs are not valid viewpoints. You may think this is only a game for science. That religion has no right to be here. Then allow me to remind you. The principle of increasing entropy[3] brings the premise and the question to the table. Are you prepared to probe deeper?

Discover the evidence that reveals *"why you should expect to live forever!"*

3

WHAT EXISTED BEFORE THE UNIVERSE BEGAN?

For sure, something caused the universe to begin! Logic-based on "cause and effect" brings us to this deduction. Everyone agrees on this idea, from Atheists to the most spiritual person on the planet. There are three possible ways for the source of the universe.

- A Supreme Being, who is personal[1].
- An eternal source that is not personal.
- A natural universe that lasts forever. (Atheism requires a "natural universe" that will exist forever, which is the same as "a natural domain that is equal to God!")

Also, there are two sources of information to help us understand. It is essential to define these as follows.

- General Revelation: This means, "What can we learn from the study of the universe?" Will science help us discern the source of the world? For Atheism, this is the only means for gaining insight into what caused the universe. However, this is relevant to the other viewpoints.

- <u>Special Revelation</u>: This is what we can know about the Source given by prophets. For example, people with miracles, dreams, visions, and the written Word of God. For Christians, the leading source is Jesus, whose life includes wonders and prophecies. In support, chapter nineteen analyzes a prophecy of Jesus that came true in our time. For Muslims, the leading source is Mohammad. Each viewpoint is probed in this book as well as in the continuing book series.

For sure, it is realistic to use "what we know" to surmise this question. "What is ultimate reality?" We must begin with general revelation, which is the only basis for the Atheist worldview.

4

ATHEISM: ITS QUEST FOR AN ETERNAL NATURAL UNIVERSE

WHAT DO ATHEISTS BELIEVE ABOUT ULTIMATE REALITY?

For sure, Atheists choose to reject the spiritual realm and the supernatural. When you die, your life ends. According to Atheism, only the natural domain exists! If true, the natural cosmos has to be eternal. This Atheist "belief" becomes a massive problem for Atheism. Because known science reveals that a natural universe will never last forever. WHY?

We have already reviewed the statement by NASA scientist Michael Pauken.

> *"I hope you don't get depressed when I tell you that the universe is going to end one day because of the 'increase in entropy' principle."*

Atheist thinkers want you to believe in a natural source that created our universe. In truth, this requires enormous amounts of faith that a natural source is equal to God. To do this, Atheism chooses to ignore known science, believing in theory only. Here is the easiest way to ponder the Atheist setback.

Think about your ancestors. Your heritage comes from your father and mother. Before that, you have grandfathers and grandmothers. Perhaps you remember your great grandfathers and great grandmothers. Each generation into the past must go on forever. If you spent your entire life saying, "great" over and over again, it would not be enough to reach into eternity past. Of course, the last thing out of your mouth would be grandfather/grandmother. Hence, your attempt to describe the source of the universe until your final breath falls short of eternity.

In fact, even if every person that ever lived on the earth did as described in the previous paragraph, it would fall short of eternity past. Stated plainly and clearly, the natural universe or sequences of them must be eternal.

For this natural domain to be eternal, the UNKNOWN universe must be infinite. Do we have any data to support an UNKNOWN world to be infinite?

No data exists! Atheism requires blind faith in weak assumptions.

In contrast with the Atheist worldview, we know that our universe will end. The world in which we live is finite.

Known science permits only one logical conclusion. A natural universe will never be eternal.

Then, how do Atheists explain this problem? Below are

overviews of why this is a massive problem for the Atheist worldview.

The Static Universe Paradigm[1]

Scientists used to believe the universe would last forever. People called this the "static universe" because they perceived the cosmos as being stable. At one time, Albert Einstein thought the cosmos had to be eternal. Then he changed his mind[2] in 1931. Why?

Scientists collected enough data by 1930 to show the universe is getting bigger. A static cosmos would not be increasing its size. Scientists learned of their false paradigm that they accepted for thousands of years.

I do not want to waste your time by writing about the details of a static universe. The most important thing is to understand the expanding universe, which reveals flaws in the Atheist worldview. At this time, we will learn three ideas that inform us of the defects in Atheist thought.

Problem No. 1: Expanding Universe

I am sure you know about the Hubble Space Telescope, named after Edwin Hubble. He used a large telescope to measure how fast the stars were moving away from the Earth. Hubble found some stars moving away at speeds up to 1,200 miles per second. He did this study in the 1920s.

Then, Edwin Hubble made a plot to help find out how fast stars were moving away from earth. We call this the Hubble constant. If we know how far a star is from the Earth, we can get a close estimate of how fast that star is moving away from us. You may be interested in learning the answer to this question. *"How fast can the most distant stars be moving away from the Earth?"*

Advanced studies reveal the answer. Most stars in the universe are moving away from the Earth at speeds faster than the speed of light. Perhaps you will be fascinated by the idea that the fastest stars are moving away from us at more than three times the speed of light!

What does this reveal about the Atheist worldview?

For sure, we will never see the light of those distant stars. No one on the Earth will ever see them, forever. They exist in our universe, but we cannot see them! When we dig a bit deeper into Atheism, this becomes important.

Problem No. 2: Created Universe

Just after World War II, some scientists put forward a new way of understanding the universe. Today, we call this the Big Bang Theory. If the Big Bang theory were right, scientists thought there would be a flash of light from the early universe. How would we find this evidence?

Two men found this evidence by accident. We call this the Cosmic Microwave Background (CMB).

At the beginning of the universe, there was no light. Scientists teach that light appeared before the stars formed. We call the first light in our world, the CMB. Tiny particles came together to produce atoms. When the atoms fused, a flash of light came forth.

Since then, we have studied the CMB many times. Scientific

instruments have confirmed the CMB. Some of these instruments circle the earth on satellites and have collected data for many years. The scientists who have studied the CMB data now realize that the static universe is false. There is truly nothing wrong with saying. "Something created the universe!"

Oscillating Universe

Atheists do not appreciate the idea that God created the universe. However, the Big Bang theory strongly supports the biblical texts of a created world. To Atheists, the concept of a Creator is unacceptable. Atheists hope to prove that God does not exist. As a result, most Atheists do not like the Big Bang Theory. Hence, they attempt to find other ways of explaining our universe.

One of the ideas is known as the "oscillating universe theory." The word oscillating means to cycle back and forth. This idea has the world begin with a Big Bang event. After the universe expands to a limited distance, it then contracts back to a small point. Then another Big Bang begins the cycle again. People thought this was possible many decades ago. An unexpected discovery showed this idea to be false. Do you remember that the most distant stars are moving away at three times the speed of light?

As time passes, those distant stars will be moving away from us at ever faster speeds. The new finding reveals that the cosmos is increasing its rate of expansion. In fact, the data related to the expanding universe supports that the cosmos will never contract.

The simplest way of understanding this idea is to take a ride in a vehicle. If you like to race cars, you know that stamping on the gas pedal will cause the automobile to increase its speed. The same applies to the universe. However, the floorboard in your car and its weak engine limits your speed. In contrast, the cosmos has no limit to accelerating. It will increase its pace of expansion forever. Scientists released this information to the public in 1998.

This idea upset a lot of scientists. They thought the data would show the universe is getting smaller (*contracting, the opposite of expanding*). Also, two scientific teams did this study. They were surprised that their data agreed.

I delight in hearing scientists when they express ideas as feelings. A man named Brian P. Schmidt was a leader of a scientific team. I genuinely like this NOVA quote from Brian P. Schmidt. Please remember, this statement is from a scientist. Distinguished Professor Brian Schmidt said, "*This universe is weird, it's creepy!*"[3]

Known science now includes the fact that the universe continues to expand at ever faster rates. However, some scientists hold onto the idea that the cosmos will contract. These are people who like to bang their heads against the wall. Because other science not mentioned herein shows, the back and forth cosmos to be false.

The Multiverse

Atheist people like to refer to themselves as scientific atheists[4]. However, the information reveals that they ignore known science because they cannot answer, "*How did the universe begin?*" Or better stated, "*How can a natural universe be immortal?*"

Let us consider their viewpoint. Do you know what is meant by the word, **multiverse**?

The multiverse calls for us to believe in a universe we can never see. Moreover, the unseen cosmos is equal to God and creates other worlds. Atheists teach that the multiverse creates so many universes that we would never be able to count them.

Are you familiar with Stephen Hawking (8 January 1942 – 14 March 2018)? People think he is just like Albert Einstein, a man with a brilliant mind. He came to believe that there is no God. In Hawking's viewpoint, the cosmos created itself. He thought that matter has always existed in an infinite universe. Do we have real scientific data to support this viewpoint?

The correct answer is that no data exists. Also, it is highly unlikely that any humans will ever observe such data. Consider what other scientists write about Hawking's viewpoint.

Dr. John Lennox wrote God and Stephen Hawking: Whose Design is it Anyway? Do you know what "circular reasoning" means?

Everyone who uses circular reasoning does the following.

- The person begins with an idea of what is already believed and accepted, which is their paradigm.
- The person must ask, "What can I think of to make this real?"

Dr. Lennox reveals that Stephen Hawking[5] uses circular reasoning. *"Did the universe self-create itself?"*

Known science does not support the Atheist worldview. We know that the universe is dying. Even if there is an unseen natural universe, it, too, is going to die. There is absolutely no data to support Hawking's viewpoint.

Let us ponder the Hawking logic. Hawking changed his view of physics in 1979. He quit using math-based proofs[6]. Instead, he accepted what he could imagine. Hawking has no data to verify his views. If you agree with him, you accept imagination only.

An excellent read on scientific atheism comes from Dr. David Berlinski. He wrote <u>The Devil's Delusion: Atheism and its Scientific Pretensions</u>. You may think of this author as a religious person. However, he is an agnostic. The National Review gives the following review.

> "An incendiary and uproarious work of learned polemical writing, unique in its scientific sophistication and authority. Rather than criticizing science from the outside, Berlinski excoriates its atheist pretensions from within."

Dr. Berlinski refers to the running down of the universe. He cites the principle of increasing entropy[7]. Also, fine-tuned laws of nature exist, which causes Dr. Berlinski to lean toward the idea of a Creator.

He quotes two well-known scientists[8]. A quick paraphrase reveals that they perceive a super-intellect created the universe. Why?

If only one of the known laws of science were slightly different, life on earth would end. This evidence points toward a Creator, who has fine-tuned the only known universe.

In contrast, consider that Atheists love the idea of the multiverse. Why?

Because in their minds, this eliminates a Creator. However, that would require getting rid of known science. Here is a breakdown of the multiverse.

- It is a natural universe.
- You can imagine anything to be valid.
- You can ignore natural law.
- It creates worlds of all kinds, and we could never count all of them!
- Anything is possible in this natural universe!

In one of the universes, Elvis Presley is alive and well, which is the extreme view of people who accept the multiverse. The vast number of worlds permits anything to be possible, even if it is not possible. The best way to explain the multiverse is by hyperbole.

Imagine that you will get married at the age of Nine hundred twenty-five years. Then you will live another thousand years in marital bliss. Maybe you will join the army and move to Fort Bliss! To the scientific atheist, this is possible. The multiverse lets all things happen naturally.

I am sure this will bring disdain. Regardless, the hyperbole I used above is a good description of the multiverse. A scientific atheist believes in a natural universe that will never run down. They have chosen to discard known science. There are heaps of problems with the multiverse.

- Known science means that natural universes only run

down.
- The multiverse is a natural universe
- The multiverse would only run down.
- The "Magic-Verse" is the correct name of the multiverse.

What is Magic-Verse?

I have been on Atheist websites to discuss their viewpoints. Since I am a Christ-follower, they would often say that I believed in magic. They equated God and magic as being equal. Let us think about the word MAGIC.

Atheism requires the word "magic" based on known science. We know that our universe is dying, which would be true of every natural world. A natural universe that creates another cosmos would indeed be magical. Why? Because natural universes only run down. Known science does not support the multiverse.

In contrast, an "all-knowing" and "all-powerful" Being could create a universe without using magic. Because the power to do it would exist in the supernatural realm.

Based on the above, the multiverse requires magic. In truth, the word "magic" only belongs in the Atheist dictionary. Magic must happen for Atheism to be valid!

Therefore, "Magic-Verse" is the true meaning of the word multiverse. Please take note that I capitalized, Magic-Verse. Why?

Atheist people are required to believe, by blind faith, that there is a natural universe that exists forever and is like God Almighty! Essentially, Atheism points to the supernatural realm. Therefore, it is credible to capitalize "Magic-Verse!"

To support this view, consider what Dr. Paul Davies wrote about the multiverse?

> "... how is the existence of the other universes to be tested? To be sure, all cosmologists accept that there are some regions of the universe that lie beyond the reach of our telescopes, but somewhere on the slippery slope between that and the idea that there is an infinite number of universes, credibility reaches a limit. As one slips down that slope, more and more must be accepted on faith, and less and less is open to scientific verification.
>
> Extreme multiverse explanations are, therefore, reminiscent of theological discussions. Indeed, invoking an infinity of unseen universes to explain the unusual features of the one we do see is just as ad hoc as invoking an unseen Creator. The multiverse theory may be dressed up in scientific language, but in essence, it requires the same leap of faith"[9].

Atheists are trying to explain the universe by appealing to unknown worlds. Those worlds exist outside the known universe.

You may choose to listen to a YouTube interview with Paul Davies. Search using the words, "Paul Davies Multiverse." In this video, Professor Davies brings up the most critical issue.

> "*The simplest thing is to try to explain the universe we see from entirely within it* without appealing to hypothetical entities

that exist outside of the universe. The problem with the multiverse, despite it being a reasonable scientific idea from particle physics, is how are we going to know about these other universes. They are not directly observable."

Perhaps you remember that the most distant stars are moving away at three times the speed of light. The distance where stars move away at light speed is about fifteen billion light-years[10]. Let us dream that we can take a photo of the whole universe. We would only be able to view stars out to fifteen billion light-years. Now an eye-opening question. How much of the cosmos would we see?

We would see less than one percent of the universe in which we live[11].

Light from those distant stars will never reach the Earth. We can see only a small portion of the known universe. How will we ever know if there is even one other universe?

What does this mean?

It is a fact that we will never observe even one more universe. Paul Davies is correct. We will never see the entirety of the only known world. To believe in the multiverse requires blind faith.

Scientific Atheism Faith Statement: "Multiverse-of-the-Gaps"

Atheists often refer to an idea called "God-of-the-gaps," which is their key phrase about people with faith in God. What does the phrase "God-of-the-gaps" mean?

If you cannot explain something because of a gap in knowledge, give credit to God. God justifies what you cannot know.

Scientific atheism has the same problem. Just replace the word "God" in that phrase with the word "multiverse." Then you have the phrase "Multiverse-of-the-gaps." In truth, scientific atheism truly is magical, based on known science!

Flawed Premises of Scientific Atheism

How can we begin to understand science and atheism?

Science, as taught in the classroom, requires that scientific inquiry cannot include the idea of God. You must assume there is no supernatural realm. However, Atheist Richard Dawkins's viewpoint is murky concerning God.

Since this book is about "ultimate reality," we must ponder it. Therefore, we must accept the assumption of a supernatural realm.

We will find that Richard Dawkins uses foggy logic regarding God. We will also learn that The God Delusion is truly about Pantheism.

To begin, think about the size of the universe. Consider how complex and inspiring it is. Think about the stars we see. You must know that as time passes, every light source in the heavens will become pitch black. The universe is doomed.

Now, ponder the premise of Atheist Richard Dawkins.

The universe is improbable!

Everyone will agree with this initial premise. Because we know the universe will run down.

Let us assume that Atheism is true, which means that the source of our universe must be from a previous world. Are there natural things that do not run down? For example, particles that are the building blocks of atoms. Atheists want us to believe that those building blocks do not run down. Is that true?

Known science reveals that those particles run down too. Every quantum particle will come to its end[12].

It is now time for the tricky question that a truth-seeking Atheist must answer. Dear Atheist friend, logic asks, *"How many universes have existed before the universe in which we live?"*

The answer is simple. An infinite number of natural worlds must go back forever. The natural universe must be equal to God!

Now back to Richard Dawkins and atheist thought. Dawkins thinks that scientific principles apply to God. His mind has jumped off logic to an emotional appeal. Dawkins achieves this by asking, *"Who created God?"*

Dawkins is asking the question based on the existence of only natural worlds. Dawkins asks a question about a so-called natural-god[13]. If Dawkins's question is accurate, it would only apply to religions that believe God is the Universe. Dawkins writes about Pantheism without knowing he is doing that.

Dawkins Evolution Solution

Based on known science, Dawkins knows that the universe is not possible. Regardless, the cosmos exists. It is vital to grasp the biggest problem for atheists. They do not know, *"How did the universe begin?"*

This problem is more significant than their belief that evolution explains life on earth. I have listened to atheists offer their solution. They believe that evolution explains everything. Then they ignore, *"How did the universe begin?"*

Here is the root issue for Atheism. **Evolution is pointless without a universe.**

The atheist answer to this problem is the "multiverse-of-the-gaps" argument. Blind faith is their footing. In truth, this is their "Magic-Verse-of-the-gaps" argument. Magic is at the foundation of atheism.

Let us rank the importance of explaining life on earth.

Rank one: Explain how the universe began. Why is it rank one? Because "evolution is pointless without a cosmos!"

Rank two: Evolution must explain life on earth. To do this, please show me data that proves nonliving matter can assemble itself into a simple living cell! No data exists[14]. Atheist people assume this to be true.

Rank three: Evolution explains life! In the scientific atheist mind, this explains everything. They choose to ignore the rank one issue completely. Scientists are attempting to find scientific answers for the rank two-issue above. Atheists must use imagination only in their attempts to find a solution for the rank one issue. At this time, it is essential to know that even if evolution could explain life on Earth, they will never overcome the rank one issue above.

From the ranking above, we deduce that evolution is a frail third rank answer to explain life on the earth. Are you willing to have blind faith in Dawkins's view without data? Even if tests could prove the rank two and three levels, it still falls short of explaining life on earth. Very simply, *"Evolution is irrelevant without a universe!"*

For Richard Dawkins, evolution is the answer. Regardless, to follow this group requires blind faith.

Conclusions about a Natural Eternal Universe

No one has found a way to prove the universe is eternal. Every attempt to find this has failed. So far, their best idea is the multiverse! In truth, that word should change to "Magic-Verse!" Will we ever see even one other world?

What do these failed attempts mean?

What does this mean for people with spiritual beliefs?

My research suggests inspiring change is coming. We are about to see an increase in knowledge that ultimate reality is spiritual[15].

5

PANTHEISM:

ITS QUEST FOR AN ETERNAL SUPERNATURAL UNIVERSE

Do you believe that the Universe is God? Consider another way of asking this question. Do you think the creation is God?

People who believe this are in a category of religions that we call Pantheism. Think about what this means. If the physical Universe[1] is God, then you and I are also a part of God.

You may also need to know that these religions think there are lots of gods. The word polytheism means many gods.

Sometimes, people who are atheists use reason and come to believe in Pantheism. After all, atheism requires you to think the universe is immortal.

If God is the Universe, then how long must the cosmos exist?

The correct answer is that the universe must last forever. In contrast, the problem of the dying cosmos defies this belief system. What is lacking for Pantheistic religions?

Explaining the Natural Universe and its Source

Dr. Paul Davies believed in natural theology, which suggests that humans can deduce that God exists as well as the nature of God from observing the natural world. He did not accept traditional religions. Paul Davies was using his views to devalue the credibility of the multiverse. His viewpoint is also relevant to Pantheism. Why? Because of the need for an eternal universe with gods of all kinds and types.

Is Paul Davies correct that the cosmos reveals stuff about the Creator? Below is a quote from Paul Davies[2]:

> "The simplest thing is to try to explain the universe we see from entirely within it without appealing to hypothetical entities that exist outside of the universe."

Paul Davies wants data of some type to explain the Creator. I agree with Paul Davies. The only difference between Paul Davies and myself comes from the following question.

Did God reveal "truth" to people living thousands of years ago?

- If yes, then we would expect God to reveal spiritual truth to people living nowadays.
- If God did not speak to people long ago, then we would not expect to hear from God nowadays. In this case, we would know absolutely nothing about the spiritual realm except by human reason, which is bound to include human error.

Let us accept the idea that God does speak to people. If true, then one of the old religions could be true. Therefore, we would have a credible source to learn about the Creator. At this point, I think it is essential that you know I chose to ignore the second bullet above. My decision comes from successfully using science to find precise dates of the time-based prophecy[3]. That information strongly supports the ancient texts foretold of specific future events to occur on an exact date, centuries into the future.

In contrast, Paul Davies would take the position that God does not reveal himself. What if we have data that refutes Paul Davies's viewpoint?

The Dying Universe: Pantheism as Ultimate Reality?

Let me now reiterate. We know that the natural universe is dying. In contrast, Pantheist beliefs require the "Universe" to last forever. I capitalized the word "Universe" because it must be immortal, which leads to asking two crucial questions.

Is the Universe God? Or did a God that transcends the universe create it?

These two questions show two categories of God.

Atheism requires magic and cannot answer the question, "How did the universe begin?"

For Pantheism to be the truth, the Universe must last forever. However, known science reveals the cosmos will not last forever. What does this tell us about the Creator?

I have read the literature of Pantheist scientists. Their solution is that we cannot observe the physical universe in its entirety. I have already stated that this is true. Why would a Pantheist scientist think this way?

If we cannot see the entire universe, then there is much more to the cosmos. Perhaps the cosmos is eternal.

In contrast, we have data that shows the cosmos will not last forever. The data comes from the study of physics. Also, what NASA people teach about the beginning of the universe. We know that the world began to exist about 13.8 billion years ago.[4] I would expect you to ask, "Show me the data!"

Perhaps some people do not know what is meant by the word "deuterium?"

Deuterium is a unique form of hydrogen, the simplest of atoms. I have created a graphic to help explain deuterium. Each atom consists of protons, neutrons, and electrons, which spin around the center.

The simplest hydrogen atom is on the left side of this graph. You will notice that there is only one proton at the center and one electron spinning around it. There are no neutrons in the basic hydrogen atom.

Deuterium is in the middle of the graph. You will notice that at the center of this atom, a neutron, was added. Deuterium consists of one proton, one neutron, and one electron. Scientists use deuterium to make the hydrogen bomb, which can destroy an entire city in an instant.

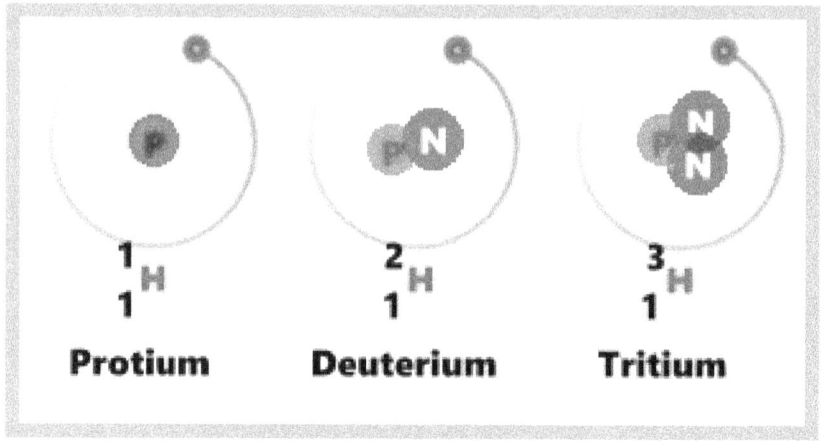

The Hydrogen Atom and its Isotopes

What does this have to do with the age of the universe?

Now the critical question. At what temperature can we create deuterium?

Will the hottest star in the universe create deuterium?

The temperature is so high that the hottest stars cannot make deuterium. In fact, stars destroy deuterium. Stars combine deuterium with other atoms to release energy.

Only the high temperature reached during creation will make deuterium[5]. In other words, this isotope reveals that God created the universe. A universe with a beginning is not eternal because the cosmos began to exist! The world began, and it will end.

Consider what people who study the early universe will tell you. They say that when the cosmos began, the high temperature made deuterium. However, this high temperature only lasted for about twenty minutes. As the temperature fell, the creation of deuterium stopped. The amount of deuterium in the universe reached its maximum level.

This information shows that the cosmos had a beginning. Deuterium verifies that the universe is not eternal. This fact points out that Pantheistic beliefs have a weak footing.

We learn two facts that oppose Pantheism.

1. Deuterium helps us to discern that the universe began to exist. We perceive that the world is not eternal.
2. We know that the universe cannot sustain itself.

These two facts support a Creator made the universe. There is no other scientific answer based on data. Also, we learned in the previous chapter that Atheism is not credible. Perhaps you do not want to accept this fact. Then you are likely using blind faith without data, which is your paradigm.

Known science points to a God that transcends the universe. Why?

1. Because the cosmos is not eternal.
2. The fact that the <u>Source</u> of the universe <u>must be timeless</u>.
3. The principle of increasing entropy[6] supports items one and two above.

Perhaps you need to think about the points above. I would encourage you to do that.

Will we find a way to verify the existence of a transcendent Being?

We have data that supports the existence of the spiritual realm.

The data comes from "The Psychic Bird Story!" I write of this freaky Twilight Zone real-life story in book four in the Expect to Live Forever book series, <u>Secrets - never heard until now - of the Spiritual Realm</u>. If you want to open your mind to what we do not see, your interest will spike. I will disclose more on this viewpoint with additional books in the continuing Expect to Live Forever book series. This book series reveals data to support why you should expect to live forever.

For observers of the Psychic Bird Story, there is only one chance in 10,000 the bird fooled us by being upset of itself. From this event, we can calculate the spiritual realm as reality is close to one hundred percent. Data analysis supports there is only one chance in a thousand, billion, billion universes that the spiritual realm does not exist.

I take the position that we can discern that God transcends time. The evidence comes from scientific dating Daniel's "time-oriented" prophecy. This method would agree with Paul Davies's statement. "The simplest thing is to try to explain the universe we see from entirely within it…"

Conclusions for Pantheistic Belief Systems

Pantheism includes people of faith. However, two facts counter that worldview.

1. Deuterium helps us to discern that the universe began to exist.
2. We know that the world cannot sustain itself due to the principle of increasing entropy.

Known science uncovers deep-seated problems for Pantheism.

For Pantheism to be valid, the universe must be eternal. The multiverse is the only possibility for that viewpoint, but the multiverse requires blind faith. We cannot observe but a small portion of the known universe in which we exist.

We are now ready to consider if God transcends the universe.

Before moving into that viewpoint, you must know one thing. It is a fact that our dying universe supports that God transcends both time and space. Why?

The transcendent Being creates the universe and lets it die! Consider a Bible verse that foretells of another creation, recorded on paper over 1,900 years ago.

"... I saw a new heaven and a new earth; for the first heaven and the first earth passed away..." (Rev. 21:1).

6

MONOTHEISM:

DID A TRANSCENDENT BEING CREATE THE UNIVERSE?

WE HAVE LEARNED THAT ATHEISM REQUIRES THE NATURAL UNIVERSE to be equal to God. That known science and Atheism point us back to the supernatural realm. Fundamentally, atheists believe in a Magic-verse[1]! Because the principle of increasing entropy causes every natural thing to run down.

Based on the above ideas and data, I created a credible premise. Known science confirms this premise.

The Premise

The natural universe cannot exist without a Supernatural source!

Using the Premise to Analyze Monotheism

DO WE HAVE ANY DATA TO VALIDATE THAT MONOTHEISM IS TRUE OR false?

The premise above logically leads to the following fundamental question.

> **Fundamental Question**
>
> **Should we expect to observe supernatural events (miracles) or the influence of spiritual beings on events in this universe?**

The logical answer is YES!

However, the logic requires us to discern otherworldly things. How would we know that a spiritual being has invaded your life to influence human events?

Consider the two methods described below.

- Method 1: Use science to find the exact dates of events foretold in Daniel's "time-oriented" prophecy. This method reveals that the texts predicted specific events would occur on an exact date, many centuries before they happened. No one can do that except for a Being that transcends time.

- Method 2: Statistical analysis of the Psychic Bird story: Everyone in the room watched this demon influenced event. This freaky story from the Twilight Zone teaches that demonic spirits influence people. The data strongly suggests that the Bible is credible. You will begin to understand my perspective as you read the book series on these profound events. Five eyewitnesses in that room observed the bird. The observations give us data to probe the spiritual realm and to derive credible conclusions about that domain.

Our created universe and the universe of the Transcendent Being

Big Bang theory teaches about the creation of time, space, and matter from nothing. Hence, that time, space, and matter did not exist before the creation of our cosmos. Big Bang theorists have taught this viewpoint for decades. How close is this viewpoint to the storyline in Genesis?

The initial verse of the Bible teaches us about a spiritual universe. Before the Big Bang event, space-time and matter did not exist. The Big Bang viewpoint aligns with the Genesis texts. Consider the ideas packed inside verse one of Genesis.

"In the beginning (time) God created the heavens and the earth (space and matter)" (Gen. 1:1).

Consider the habitation in which the Creator exists.

- The first phase, "In the beginning," reveals the Source exists where time does not exist. Hence, the Source is eternal and will transcend time!
- The following expression, "… created the heavens …" suggests a place where space does not exist. As humans,

we may not understand a place where there would be no space. However, we can grasp that the Creator will transcend space.

> **NOTE:** The initial verse of the Bible reveals the Creator as a Being that transcends space-time. Hence, knowledge of all future events as well as when they happen. We can perceive that the Creator exists outside the limits of space and time.

- The completion of the second catchphrase, "…created the heavens and the earth …" reveals a place where matter does not exist. Therefore, a place where only spiritual beings live. Our physical bodies could never enter that spiritual realm.

The initial verse of the Bible teaches us to believe in an entirely spiritual universe[2]. God exists in an everlasting supernatural realm, which might be the place we will live forever.

Other scriptures support this viewpoint.

> "By faith we understand that the universe was formed at God's command, so that what is seen was not made out of what was visible" (Hebrews 11:3 - NIV).
>
> "For thus says the LORD, who created the heavens (He is the God who formed the earth and made it, He established it and did not create it a waste place, but formed it to be

inhabited), "I am the LORD, and there is none else" (Isa. 45:18).

"... for You created all things, and because of Your will they existed, and were created" (Rev. 4:11 – NASB95).

These biblical texts show the universe created from nothing. Big Bang theory lines up with biblical texts that the creation came from nothing. In contrast, Pantheism teaches that the gods come from what already exists.

Time Travel and the Impossibility of Foretelling the Future

Humans cannot travel faster than the speed of light. If we could, then we could go to the past with our knowledge of events that have already occurred. For example, what could I do if I went back in time to 1930?

I would know who would win every future U.S. Presidential election. I would know what investments to make to become super-wealthy. I would invest heavily in IBM or Apple Computer. The story goes on and on about predicting the future.

If traveling back in time were possible, then we would be scientific prophets. We would know the details of many future events, including the exact date of the incidents.

The Crossing, a TV series, brings clarity to this viewpoint. In the storyline, a historian 180 years from the future knows the specific time and place of an earthquake. It is a great science fiction story. This movie is an excellent example related to finding the precise dates of predicted events in Daniel's "time-oriented" prophecy! The texts in Daniel foretell of specific incidents that will occur on an exact date, recorded on paper centuries before.

A Supreme Being that transcends time would know the details of specific future events. Were the texts given so we would only know that God can foretell the future?

The answer is no[3].

Consider that the foretold events bring clarity to our spiritual nature. For example, the burning down of Herod's Temple occurs on the final 14,000th day in the time-based texts. Since the temple was for sacrificial services, what does its destruction mean?

For sure, Herod's Temple existed only for spiritual purposes. In the time-based prophecy of Daniel, what occurs on the final 14,000th day is meaningful. Due to the event happening on an exact date, it appears that God-like control took place to cause the predicted event to occur on that very date. There seems to be divine influence over various people, and those people did not know of this influence. In my opinion, this is a credible idea based on the research.

Why would God do this?

In my estimation, this high-level control reveals spiritual truth. Are you a spiritual seeker? Do you want to know that we are spiritual beings at our core[4]?

It is certain that humans cannot foretell of specific future events to occur on an exact date, centuries into the future. Neither can other created beings foretell the future. For example, can aliens from outer space (*to include what many believe to be factual*) predict the future?

The beings cited above will never be able to foretell the future. Because time travel to the past is impossible for created life forms. What would it mean to find texts that predict a future specific event on an exact date, many centuries before the event occurs?

Let us ask a different series of questions from the viewpoint of the supernatural realm.

- What if the foretold event reveals spiritual ideas?
- Does that specific event show that we are spiritual at our core?

If we answer "Yes" to these questions, then we have uncovered evidence of a God that transcends time. Finding texts that are this accurate is the "key" to verifying spiritual truth. Here are two verses that support this method as a way of finding spiritual truth.

"... Declaring the end from the beginning and from ancient times things which have not been done, Saying, 'My purpose will be established, And I will accomplish all My good pleasure'" (Isa. 46:10).

"Let them bring forth and declare to us what is going to take place; As for the former events, declare what they were, that we may consider them, and know their outcome; Or announce to us what is coming" (Isa. 41:22).

I like both texts. However, Isaiah 41:22 contains a phrase that is relevant to finding precise dates in the "time-oriented" writings about the future. Because we are looking at events of the past to find out if the texts did foretell the future. "... *declare what they were*" means to consider what occurred in the past. In this case, the passages reveal that they predicted the future. The end of this verse proclaims, "... *announce to us what is coming.*"

Logic and science are the tools used to find evidence of a transcendent Being. For sure, a God that transcends time knows the future. The precision of this prophecy strongly suggests that other biblical texts reveal even more about the spiritual realm. The details from this research support the influence of God over human events related to burning down both Solomon's Temple and Herod's Temple. Are you ready to grasp that you should expect to live forever?

7

USING LOGIC TO PROBE

BIBLICAL TEXTS FOR SPECIFIC FUTURE EVENTS

To gain insight into texts that foretell the future, we must use both science and the math used in the passages. If you think you can ignore biblical math, then you will never find the true meaning of these texts. I have outlined the approach so you can follow the logic and the required math. You will learn that biblical math gives us repeatability.

To begin, how do we know that the texts foretold the future?

We must show evidence that the texts preceded the foretold events. Hence, we must know the date when Daniel wrote these time-based texts about the future.

Then I will give a general outline of what events the texts foretold. What incident happened first, then the ones that followed in sequence.

I have learned from this research that we must know the day of the week upon which each incident occurred. To achieve this, we have eyewitness accounts regarding the burning down of

Herod's Temple. Ancient sources document that the temple burned down on a Sunday. For the other events, we must use biblical math in the time-based prophecy. We learn that all time-based texts use the number seven. Since there are seven days in a week, we logically conclude that every future event must occur on a Sunday.

Can we trust the biblical math?

We will find the numbers used in the Bible give us repeatability. For example, the number 360 is the constant that always unveils 14,000 days. I know that some people do not like using the number 360. Regardless, the mystical 14,000 days period is present below the surface of biblical texts at the time of Moses and Joshua. Also, this repeats at the time of Jesus. For sure, the number 360 always reveals the 14,000 days, which shows repeatability.

After this, we begin to delve into the time-based texts that foretell the future.

We find that temple destruction events appear to be controlled by God. For example, why were Solomon's Temple and Herod's Temple burned down on the final 14,000th day?

We learn about a third temple. It appears that a heavenly temple, not of this world, replaced the earthly temples. In particular, we surmise that Jesus' death on the cross did away with the earthly temples on the 14,000th day. Why?

There is a spiritual purpose for the time-based texts about future events. Is Jesus the ultimate sacrifice?

Does the burning down of Herod's Temple on the 14,000th day validate Jesus' death and resurrection?

To support that this is the correct conclusion, burning down Herod's Temple is one of the specific events foretold to occur on an exact date from many centuries before.

Finally, we discover evidence that validates the research of Sir Robert Anderson. Modern-day scholars consistently reject Anderson's conclusions. However, they always laud him for making this sophisticated prophecy easy to understand.

My research into ancient astronomy reveals that Sir Robert Anderson was correct. Unveiling the 14,000 days that begin with Moses and Joshua, then end in the twenty-first century, are related to Sir Robert Anderson's research.

8

WHEN WAS DANIEL WRITTEN?

WE FIND THE TIME-BASED PROPHECY IN THE NINTH CHAPTER OF Daniel. The passages begin with his prayer about the future of the Jewish people. As Daniel prayed, the angel Gabriel appeared. That is when Daniel received the time-based texts about the future, which happened about 538 BC.

Imagine that a human speaks with a spiritual being! Then the angel predicts the future with time-based targets. Because of the time elements, we need to know when Daniel recorded the texts on paper.

Two types of people get involved with this decision on the composition date for Daniel.

- Conservatives who accept the supernatural
- Rationalists who reject the supernatural

JOHN ZACHARY

Comparing Daniel to the Dead Sea Scrolls Writings

Professor Gleason L. Archer writes about the word usage in the book of Daniel. How do those words in Daniel match up with the word usage in the Dead Sea Scrolls documents, written from 250 to 100 BC?

This data supports that Daniel wrote the book in the late sixth century BC or a short time later. Archer thinks it happened in 530 BC based on the fact that language changes occur as time moves forward.

To illustrate this idea, *"Do you know how much the English language changed over time?"*

Consider English from Chaucer's poems in the late 1300s, which shows how much English has changed over 600 years[1].

Word	Fourteenth Century	Twenty-first Century
Milky Way	*Which men clepeth the **Milky Wey***	*which men call the **Milky Way***
Vacation	*Whan he hadde leyser and vacacioun*	*When he had leisure and took some vacation*

In the same way, the language in Daniel (Sixth century BC) does not match the Dead Sea Scrolls era books (Third to second century BC). Here is what Professor Archer wrote about comparing Daniel to documents written centuries later in Palestine. I have paraphrased Professor Archer's writing with the actual text placed in this footnote[2]. Please read the footnote if you want to study this. Here are the conclusions.

Data shows Daniel to be centuries older than a third century BC document. Also, the language supports that Daniel wrote the texts in Babylon. In contrast, the rationalist claim that someone wrote Daniel in 165 BC in the area of Palestine. In truth, the rationalist have zero data to support their "opinion!" Data supports the 530 BC date for writing.

People who choose the 165 BC composition date avoid the data. They reject the Bible came from a God that transcends time.

There is more to this because we have other information that suggests an earlier date for Daniel. Below is a quick review of those sources.

Josephus[3] wrote about Alexander the Great. When Alexander entered Jerusalem in 332 BC, he observed the book of Daniel. Jewish priests read the texts in Daniel to Alexander. The passages foretell of a Greek that would defeat the Persian Empire. We learn from Josephus that Daniel wrote the book long before 332 BC. Consider what Josephus wrote about this meeting.

"... he (Alexander the Great) gave his hand to the high priest and, with the Jews running beside him, entered the city (Jerusalem). Then he went up to the temple, where he sacrificed to God under the direction of the high priest and showed due honor to the priests and to the high priest himself. And, when the book of Daniel was shown to him, in which he had declared that one of the Greeks would destroy the empire of the Persians, he believed himself to be the one indicated; and in his joy, he dismissed the multitude for the time being, but on the following day he summoned them again ..." (Antiquities of the Jews, 11.8.5).

Perhaps you know of the Septuagint? Ancient scholars translated the Old Testament into the Greek language around the year 250 BC. According to Josephus, the Septuagint included the book of Daniel.

From a scholar's point of view, people living closer to the events know more. We can put more trust in people living close to past historical events. In contrast, people living over 2000 years into the future are using bias for their conclusions.

More evidence for the earlier date comes from the Dead Sea Scrolls. Those people wrote two commentaries on Daniel. In one of those, they refer to the "time-oriented" prophecy about the coming "Anointed One" as follows:

"Anointed of the Spirit, of whom Daniel spoke!"

Another text from a Dead Sea Scroll[4] quotes from Daniel 12:10. The author refers to Daniel by writing, the **"book of Daniel the prophet."**

Ancient people accepted Daniel as a prophet. Since Daniel foretold of the destruction of Jerusalem and the temple, perhaps that affected Josephus. For sure, Josephus wrote an eyewitness account of Romans burning down the temple. Daniel's time-based texts foretold of that event. Did this affect how Josephus wrote?

Burning Down of Herod's Temple in AD 70

If you take the position that someone wrote Daniel in 165 BC, then the texts foretold of burning down Herod's Temple in AD 70. The predicted event occurred 235 years into the future.

However, the 530 BC date supports that Daniel foretold the burning down of Herod's Temple 600 years into the future.

Regardless of the two viewpoints, the final event did occur in the distant future.

Now consider the idea of 14,000 days hidden in the time-based texts. The research shows that Romans burned down Herod's Temple on the 14,000th day in the time-based prophecy.

Based on this information, which viewpoint does this support for the composition date of Daniel?

9

WHAT EVENTS WERE FORETOLD?

Daniel's texts reveal future events will occur as time moves forward. I list each incident in order from the first to the final one.

Biblical scholars refer to these texts as the prophecy of the seventy weeks. It is vital to know the meaning of the word weeks.

The original Hebrew word for "week" equates to a time of seven years. We often refer to this as a Sabbatical. This Sabbatical comes from Leviticus chapter twenty-five, which is the counting for Sabbath years. Please note that the number seven becomes the basis of finding the exact dates of the foretold events.

The First Event

The prophecy begins with a decree issued by a future king to restore and rebuild Jerusalem. Here is the text from the prophecy.

> "...know and discern that from the issuing of a decree to restore and rebuild Jerusalem..." (Dan. 9:25).

The Second Event

The second event is the appearance of the Messiah. Here is the text from the prophecy.

> "...until Messiah the Prince..." (Dan. 9:25).

NOTE: This time-based prophecy only gives time elements from the first event to the second event. The texts provide no other time elements from the sixty-ninth week to the seventieth week.

The Third Event

The third event reveals what will happen to the Messiah. These texts reveal the purpose of the Messiah. I have emboldened the purpose texts in the verses below.

> "...Then after the sixty-two weeks the **Messiah will be cut off** and have nothing..." (Dan. 9:25).

Verse twenty-four has the purpose of the entire prophecy, which relates to this event:

"²⁴Seventy weeks have been decreed for your people (*Jewish people*) and your holy city (*Jerusalem*), to finish the transgression, to make an end of sin, **to make atonement for iniquity**, to bring in everlasting righteousness, to seal up vision and prophecy, and to anoint the most holy place" (Dan. 9:24).

The Fourth Event

The fourth event is the destruction of Jerusalem and the temple. Here is the text from the prophecy.

"...The people of the prince who is to come will destroy the city (*Jerusalem*) and the sanctuary (*Herod's Temple*)..." (Dan. 9:26).

10

UPON WHAT DAY OF THE WEEK DID EACH EVENT OCCUR?

IN THE PREVIOUS CHAPTER, WE LEARNED THAT A PERIOD OF SEVEN years is the foundation of the time-based prophecy. The number seven reveals information about the day of the week. The prophecy foretells of events that will always occur on the same day of the week. Why?

Seven days each week are the foundation of our calendars. What day of the week are we to expect?

The answer comes from the eyewitnesses of the fourth event. On which day of the week, did Romans burn down Herod's Temple?

The Eyewitnesses

Josephus watched Herod's Temple burn to the ground. He wrote in a way that strongly suggests the event occurred on a Sunday.

Another person from that era, Rabbi Akiva, born in AD 50, was

20 years old when Roman soldiers burned down Herod's Temple in AD 70. Rabbi Akiva knew this event occurred on a Sunday. He also knew that Babylonian armies burned down Solomon's Temple on a Sunday. He taught this to his student Rabbi Yose.

Rabbi Yose wrote that both temples burned down on a Sunday in <u>Book of the Order of the World</u>. He published this book close to AD 150, about eighty years after Herod's Temple burned down.

Since we know the day of the week, we can find the exact date of each event in Daniel's time-based texts. We only need other information[1] to find the precise day upon which each event took place.

In conclusion, each event tied to the time matrix with the number seven occurred on a Sunday. In chapter nine, events one and two happen in the seven-day time matrix. The fourth event is unique because we find no math to link it to the seven-day pattern. However, we will discover a period of precisely 14,000 days between the second and fourth events. You may be intrigued to learn that the biblical constant of 360 days per year is required to give us the 14,000-day pattern. We begin to discern God-like control of human activities related directly to the burning down of Herod's Temple.

Are you prepared to learn of biblical numbers that have repeatability?

11

REPEATABILITY OF THE BIBLICAL CONSTANTS

Do you want to understand Daniel's time-based texts that foretell the future?

If you answer yes, then you must use the math inside the texts of the Bible. We will find that the number 360 days in a year to be a biblical constant that explains the time-based documents. This 360 value always results in a period of 14,000 days. The number 360 gives us repeatability.

Why is repeatability necessary?

Engineers desire repeatability to achieve reliable hardware designs. Would you step onto an aircraft if the state-of-the-art equipment did not have repeatability?

I have read of people who reject using the value of 360 in biblical prophecy. However, this 360 value gives us repeatability based on the positions of the sun, the moon, and the earth. The alignment of these heavenly bodies to 360 days per year support it to be credible[1].

Moreover, 360 is used in modern-era cultures every day. For example, we have 360 degrees in a circle. Ancient peoples began to use 360 degrees in a circle, centuries before Christ. The idea may have come from ancient peoples studying the sun to create a calendar of some type. The original Egyptian calendar contained 360 days in a year. You may be intrigued that Microsoft Excel has a function for 360 days per year.

Perhaps you would like to know where we get the biblical value of 360 days per year?

We find the 360 value in texts from the books of Genesis[2], Numbers[3], and the book of Revelation. To keep this to the point, consider the book of Revelation verses that use the value of 360 days per year.

> "[2]... they will tread underfoot the holy city for forty-two months. [3]And I will grant authority to my two witnesses, and they will prophesy for twelve hundred and sixty days" (Rev. 11:2-3).

The text above shows 30 days in a month. Divide 1,260 days by the forty-two months as follows.

$$\frac{1,260 \; Days}{42 \; Months} = 30 \; Days/Month$$

By multiplying 12 months of 30 days' length, we get a constant of 360 days per year. Since the book of Revelation is prophetic, many people refer to this biblical constant as a prophetic year.

Sir Robert Anderson used the value 360 to interpret the time-based texts. After more than four decades of research, I am grateful for Sir Robert Anderson's research. He published <u>The Coming Prince</u> in 1894. Anderson's research laid the foundation that permits us to find the 14,000 days hidden in Daniel's time-based texts.

There are two periods in the Bible where the 360 days give us the 14,000 days sequences. Consider the application hereafter.

By the way, if you do not like math whatsoever, you can skip forward to the next chapter. However, please be assured that I have a college degree in math and graduated with honors. Anyone with a calculator that follows the biblical math will get precisely 14,000 days. The guideline to get the 14,000 days surrounding the lives of Moses and Joshua requires using the value of 360 days per year.

Moses and Joshua – 14,000 Days
based on the 360 Day Biblical Constant

We will find precisely 14,000 days hidden in biblical texts when Moses and Joshua lived. This discovery is from Pastor Charlie Garrett. Many people have read these texts for thousands of years. Why is it that only Pastor Charlie found these 14,000 days in the modern era?

I have studied this initial set of 14,000 days in-depth. It turns out that this 14,000 days period is mystical. As you read into this chapter, you will find out what I mean by using the word mystical. Please consider that the word mystical can also mean supernatural, transcendent, or otherworldly.

You may be thinking, *"This cannot be for real!"*

Regardless of your initial thoughts, everyone on the planet, with a calculator, will get a value of 14,000 when you use the biblical constant of 360 days in a year.

Using the word "constant" simply means a number that is always the same. If you choose to ignore this biblical constant, then you will not get the 14,000 days. The constant of 360 days in a year sticks out like a sore thumb. Because you realize it is the basis for the value of 14,000 days hidden one layer below the surface of the biblical texts. Think of a miner digging for gold. You have to get your pick and shovel out to find these gems beneath the surface of the documents.

The 14,000 Days for Moses and Joshua

The first day for counting the 14,000 days begins on the precise day Israel left Mount Sinai. Moses had already received the Ten Commandments while they stayed at Mount Sinai. It is worth noting that the entire nation did nothing unless God led them. Therefore, God was controlling the date on which they left Mount Sinai.

> "…it came about in the second year, in the second month, on the twentieth of the month, that the cloud was lifted from over the tabernacle of the testimony" (Num. 10:11).

The passage above tells us how much time had passed since they left Egypt. It is vital to understand this includes the entire month of Nisan in which they left Egypt.

Based on the verse above, an entire year had passed, which begins the 360 days in a year idea.

According to this verse, it is in the second year. Also, the people left on the twentieth day of the second month.

That means the first month had thirty days.

And that the second month, the day of departure, it is the twentieth day.

Add the three numbers above, and we get 410 days. Perhaps the following will help you grasp the numbers[4].

360 Days + 30 Days + 20 Days = 410 Days

Now we find the day on which the Jewish people crossed the Jordan River into the Promised Land. We know that they spent forty years in the desert based on the following passage.

> "...the sons of Israel walked forty years in the wilderness..." (Joshua 5:6a).

Please remember, we are using the 360 days in a year value. So, we multiply the 40 years times the 360 days in a year to get 14,400 days.

$$360 \frac{Days}{year} \times 40\ years = 14,400\ Days$$

The next verse reveals that they crossed the Jordan River precisely ten days later.

> "^{19}...the people came up from the Jordan on the tenth of the first month and camped at Gilgal on the eastern edge of Jericho" (Joshua 4:19).

According to the biblical texts, this is the beginning of the 41st year in the first month of that year. We just add ten days to the 14,400 days to get 14,410 days.

$$14,400\ Days + 10\ Days = 14,410\ Days$$

The 14,410 days begin in Egypt and end on the day they cross into the Promised Land. Now we must subtract the days between the two calculations above, which give us the mystical value of 14,000 days hidden deep in the texts.

$$14,410\ Days - 410\ Days = 14,000\ Days$$

Why would we find 14,000 days at the time of Moses and Joshua? Why didn't somebody find this hidden period a long time ago? These are great questions. However, the essential inquiry follows.

"Why did they cross into the Promised Land on the final 14,000th day?"

Before moving forward, you need some vital information. To begin, this is the first set of 14,000 days we have found. We will find a total of four groups of 14,000-day periods. The final 14,000th day turns out to be an event with high-level spiritual meaning. Why?

Because what happens on that final day affects you and every person that will ever live. I am referring to your soul. The fact that you should expect to live forever.

For now, I will concentrate on the 14,000 days at the time of Moses and Joshua.

Because this turns out to be a mystical period, for example, you see people with your eyes. What they look like, how they act, and other external things. In contrast, the mystical would be what is in their heart. There is much more to each of us than what we see on the surface.

We have found the first set of 14,000 days that links up with the ones that follow.

We will find the second set of 14,000 days that depends on using the value of 360 days per year in Daniel's time-based texts.

We learn that we can only understand the Bible by letting the texts teach us their meaning. The 14,000 days sequences found only in the Bible reveal spiritual truth. Why? Because we learn that the periods relate to foretelling future events to an exact date. In support of this viewpoint, the activities align with the positions of the sun, the moon, and the earth.

In the next chapter, we will discover that the 14,000 days at the time of Moses and Joshua turn out to be mystical.

12

INSIGHTS FROM THE MYSTICAL 14,000 DAYS

I LIKE TO REFER TO THE FIRST PERIOD OF 14,000 DAYS AS MYSTICAL. Why?

We will find that the date for entering the Promised Land does not align with the historical events from the Exodus story. A gap in time exists. Let me explain. To do this, I want to help you understand how easy it is to count 14,000 days. Although this may sound hard to learn, you will discover it is easy to do. Think of the day you were born, your birthdate! How long does it take for you to be exactly 14,000 days old?

You may think this is a waste of time. However, there is a purpose in why I am teaching this to you. You will learn how to find 14,000-day alignments hidden in biblical texts. Also, it will help you understand what I am writing.

Easy Two-Step Process

Step 1: Add thirty-eight years to the year you were born.

Step 2: Then, count forward four months to the very day.

When you do this, the 14,000th day will be within two days of that date.

Here is an example. If you were born on Friday, April 6, 1900, do the following.

Step 1: Add thirty-eight years to the year 1900 to get the year 1938.

Step 2: Count forward four months from your birthday on April 6.

1. May 6
2. June 6
3. July 6
4. August 6

Therefore, expect August 6, 1938, to be the 14,000th day.

However, that turns out to be the 14,001st day. Why?

The reason is that the birth date on April 6, 1900, occurs on a Friday. However, the day of the week for August 6, 1938, happens on a Saturday. If you were born on a Friday, then the 14,000th day must occur on the same day of the week, which in this case, is a Friday[1].

Therefore, the 14,000th day from Friday, April 6, 1900, will occur on Friday, August 5, 1938.

Why do we need to have the same day of the week?

Because the number 14,000 is divisible by the number seven. There are seven days each week. Hence, we can grasp the same day of the week to have precisely 14,000 days.

Now, let's go back to the 14,000 days at the time of Moses and Joshua. It is time to show you why that period turns out to be mystical. We do this by adding 14,000 days to the day that the Israelites left Mount Sinai. We use the biblical text for the start of the 14,000 days.

> "...it came about in the second year, in the second month, on the twentieth of the month, that the cloud was lifted from over the tabernacle of the testimony" (Num. 10:11).

It was the second year. Add thirty-eight years to get forty years[2].

Now count forward four months from the biblical text above. The words show that it was the second month on the twentieth day when they began to travel.

1. The third month, the twentieth day
2. The fourth month, the twentieth day
3. The fifth month, the twentieth day
4. The sixth month, the twentieth day

Based on this, the Israelites would have crossed into the Promised Land in the sixth month, the twentieth day, thirty-eight years into the future.

Is that what happened?

The passage in Joshua 4:19 records the tenth day of the first month, which does not align with the sixth month, the twentieth day. Hence, we find a mystical period of 14,000 days at the time of Moses and Joshua. Regardless of the mystical nature of the 14,000 days at that time, we will find this is important as you read on into this research. We will see a deeper meaning to the Bible. Although it is deep, it is also simple.

Moreover, since the mystical period of 14,000 days exists, what does this reveal about the Bible?

For me, it means there is a deeper level for these texts. The mystical 14,000 days support that these texts are from a divine source. Support comes from the fact that we will uncover three other sets of 14,000 days hidden inside the Bible. Finally, it is vital to note that we only find the 14,000 days by using the value of 360 days in a year. The number of 360 days in a year is strictly from the Bible.

Now a critical question. Was there an actual real value of 14,000 days hidden in the writings surrounding the lives of Moses and Joshua?

There may be such a time, but it is impossible to find that precise period. I have added a footnote for people interested in that research[3].

The 14,000 days at the time of Moses and Joshua lays a foundation for the future. What can you gain from the 14,000 days sequences?

First, we learn that the events on the 14,000th day occur at crossroads.

Second, the final 14,000th-day incident reveals high-level spiritual truth. What happens on the earth reveals information that is not of this creation. Stated another way, we uncover eternal truth. At this time, you only know of the first set of 14,000 days as a mystical event related to entering the Promised Land. In my viewpoint, this mystical event symbolizes the day that you will enter heaven.

Many modern-era scholars do not accept the Bible. However, these studies confront their viewpoints. For example, scholars that study ancient Egypt claim that the exodus of Israel from Egypt did not occur. The 14,000 days hidden in biblical texts support that it did happen. That there is eternal information hidden within the passages.

What can we perceive about the initial 14,000 days?

We know that they are mystical.

Is there a symbolic meaning regarding the mystical 14,000th-day event?

The book of Revelation is based on symbolism. Entering the Promised Land appears to be symbolic of entering Eternal Life. We will find this concept to be consistent in regard to the future temple destruction events. Clearly stated, temple destruction events on their 14,000th day point us toward eternity.

13

JEWISH TEMPLES: PURPOSE AND CONTROLLED DESTRUCTION

CONSIDER THE RELATIONSHIP BETWEEN SOLOMON'S TEMPLE AND Herod's Temple. I am writing these ideas so you get a heads up on what you will find in subsequent chapters. The information strongly suggests God-like control of each temple's destruction based on the following list of factors. Four of these factors come from Rabbi Yose, who wrote <u>Book of the Order of the World</u>, in roughly AD 150.

Rabbi Yose was a student of Rabbi Akiva, who was born about AD 50. Rabbi Akiva would have been twenty years old when Romans burned down Herod's Temple in AD 70. Hence, Rabbi Yose's information comes from a person who knew the details about burning down Herod's Temple. I add two more factors to this list based on the scientific dating of time-based texts recorded in Daniel.

Rabbi Yose – Temple Destruction Factors

- Each temple burned down on the same date on the Hebrew calendar, which happened on the tenth day of the fifth Jewish month[1].
- Both temples burned down on the same day of the week, Sunday[2].
- Both temples burned down during the priestly division of Jehoiarib[3]. In 1st Chronicles 24:7-18, we find the first division of priests named Jehoiarib. We also find there are twenty-four divisions. We can glean two things from this concept.

THE FULL CYCLE OF TWENTY-FOUR DIVISIONS HAD COMPLETED THEIR service. Think of completion or the end.

Why were both temples destroyed with the first division of priests doing their service? The full cycle of twenty-four suggests change.

For the two paragraphs above, we get the idea of completion and a new beginning. When you consider full usage of all the priests and the end of the earthly temples, the New Testament refers to an eternal temple, which is not of this world. Earthly temples give way to an eternal temple!

- Both temples were destroyed immediately after the Sabbatical year[4]. Again, the suggestion of completion of time as a factor. You can also think of the seven-year sabbaticals in Daniel's "time-oriented" prophecy.

Scientific Dating Daniel's Time-Based Prophecy
Two New Bulleted Factors

- Foreign armies burned down both Solomon's Temple and Herod's Temple on the final 14,000th day.

NOTE: This will be shown in subsequent chapters.

- Daniel's prophetic texts foretold of a foreign army burning down Herod's Temple, many centuries before it happened. This prophecy supports that the Source is a Being that transcends time. The concept of 14,000-day sequences with the destruction of both temples on the final 14,000th day reveals that temple destruction events have high-level spiritual meaning.

This repeating sequence strongly supports God controlled some humans to align these events. Is there a relationship based on how and when both temples burned down?

Consider that both edifices were designed and used for sacrificial services. In each temple, the priests sacrificed animals, so your sins were removed. That animal shed its blood to pay the price for your sinful acts. As a result, you were cleansed of your sins and made righteous before God.

What if you did not use the sacrificial services?

Then your sins were not forgiven[5]. Below is a scripture related to the forgiveness of sins found in the Old Testament. This biblical concept began at the time of Moses speaking directly to God. Despite yourself and the fact that everyone sins, can you be forgiven of your sins?

> "For the life of the flesh is in the blood, and I have given it to you on the altar to make **atonement** for your souls; for it is the blood by reason of the life that makes atonement" (Lev. 17:11).

What is the meaning of atonement?

> The central message of the Bible is atonement. The meaning of the word is simply at-one-ment, for example, the state of being at one or being reconciled to God[6].

Daniel's time-based prophecy includes atonement as the purpose of the prophetic texts. We learn that the foretold events are related to sacrificial offerings with the shedding of blood. In Daniel's time-based prophecy, we find two events directly linked with atonement. Also, there are precisely 14,000 days between these two foretold events.

The first event turns out to be Jesus' entry into Jerusalem on Palm Sunday, Nisan 10, as the lamb of God. Do you remember that God directed Moses to choose a sacrificial lamb on Nisan 10[7]. The 14,000th day occurs thirty-eight years and four months into the future on Ab 10[8]. Romans burned down Herod's Temple, which ended the possibility of making sacrificial offerings at that temple. Consider how atonement in Daniel's time-based texts relates to the foretold events.

> "²⁴Seventy weeks have been decreed for your people (*Jewish people*) and your holy city (*Jerusalem*), to finish the transgression, to make an end of sin, **to make atonement for iniquity, to bring in everlasting righteousness**, to seal up vision and prophecy, and to anoint the most holy place" (Dan. 9:24).

The verse above reveals the purpose of the time-based prophetic texts. After the word atonement, the next idea is that this will bring in *everlasting righteousness*. How can this predicted future atonement bring about eternal right standing with God[9]?

New Testament documents clearly state that Jesus died on the cross, then rose from the dead. We discover that Jesus alone is the historical character that fulfilled Old Testament prophecies. We perceive that the Messiah foretold in Daniel's time-based texts is the ultimate sacrifice!

Were both Solomon's Temple and Herod's Temple burned down on the 14,000th day to validate these events as acts of God?

Consider that the book of Hebrews suggests that temple destruction events are related to the Messiah.

> "¹¹... when Christ appeared as a high priest of the good things to come, He entered through the greater and more perfect tabernacle, not made with hands, that is to say, not of this creation; ¹²and not through the blood of goats and calves, but through His own blood, He entered the holy place once for all, having obtained eternal redemption. . . .

¹⁵He is the mediator of a New Covenant" (Heb. 9:11-12, 15).

These texts portray a spiritual temple that is not of this world, where you have everlasting righteousness.

After decades of research, I conclude that the earthly temples burned down with six common factors noted at the beginning of this chapter. Those factors point us to an eternal temple that is not of this world. It will be impossible to destroy this superior and eternal temple. The purpose of all these studies reveals that you should expect to live forever!

14

BURNING DOWN SOLOMON'S TEMPLE

THE 14,000TH DAY

How do we know that foreign armies burned down Solomon's Temple on the final 14,000th day?

We learned in chapters eleven and twelve that the 14,000 days at the time of Moses and Joshua are mystical. Everyone with a calculator will get the 14,000 days answer when they use those guidelines.

It seems that the mystical 14,000 days reveal the biblical texts are from a Supreme Being, who is mystical, powerful, and all-knowing beyond what we could begin to understand.

Are the 14,000 days related to burning down Solomon's Temple mystical?

We will discover that they are both mystical as well as actual days. The mystical comes from the fact that there is no exact biblical date in the Bible for counting the first day of this 14,000 days group. I will explain this soon.

Regardless, we do find biblical evidence to support 14,000 days.

Before we begin finding the answer to the initial question, you need to know that the 14,000 days are related to the sacrificial system. To understand this idea, you need a short overview of the original sacrificial system, how it began, then how it ended.

We will also discover that this set of 14,000 days foretells of future events fulfilled by Jesus as the Messiah.

The Beginning of the Sacrificial System

At the very first Passover in Egypt, the people chose a lamb. The exact date occurred on the tenth day of the first month. This lamb was set apart for four days.

On the fourteenth day, they killed the lamb for its blood. Then the people took the blood and placed it onto the doorpost of their homes. The following biblical texts reveal the day for choosing the lamb.

> "^{2}'This month shall be the beginning of months for you; it is to be the first month of the year to you. ^{3}'Speak to all the congregation of Israel, saying, 'On the tenth of this month they are each one to take a lamb for themselves, ...'" (Ex. 12:2-3).

The important thing is that this began the sacrificial system. Four days later, they killed the lamb for its blood. Why?

Because the Pharaoh in Egypt would not let the Jewish people leave the land. Why not?

The Jewish people worked as slaves for the Egyptians. So the Egyptians could get more things done with slaves working. Before selecting the lamb, there had been nine plagues. Pharaoh would not budge in the least way on setting free the Jewish slaves. It took something drastic to change Pharaoh. In this case, the Egyptians would taste death. That is what it took to free the slaves.

Consider the reason for killing the lamb for its blood again. After the people collected the blood, they smeared it on the doorposts of their houses. After sunset, the Jewish people ate the lamb inside their homes. During the night, the destroyer came into Egypt. If a person did not have the blood on their doorposts, then the firstborn son died that night.

In contrast, the destroyer would not enter the home of the people who had the lamb's blood smeared on their doorposts. Simply stated, death *passed over* that house, which is why the holiday is called *the Passover*.

The tenth plague resulted in the firstborn son of every Egyptian family dying that night. They did not believe in Moses' instructions for placing the lamb's blood on their doorposts.

In the New Testament, the story gets repeated. Jesus entered Jerusalem on the tenth day of the first month. Four days later[1], Jesus died on the cross during that Passover week. Jesus died as the Lamb of God, who takes away the sin of the world. Isaiah wrote about the future of Jesus' death on the cross.

---

[5]"...He was pierced through for our transgressions, He was crushed for our iniquities; The chastening for our well-being fell upon Him, And by His scourging we are

healed. ⁶All of us like sheep have gone astray, Each of us has turned to his own way; But the LORD has caused the iniquity of us all to fall on Him" (Isa. 53:5-6).

In Egypt, the firstborn sons of Egypt died on the night of Passover. Amid deep sorrow, Pharaoh let the Jewish slaves go free. You may recall the movie <u>The Ten Commandments</u>. It is about the Passover and the miraculous crossing of the Red Sea.

In the desert, Moses received even more guidance on this system. You may recall the Indiana Jones movie, <u>Raiders of the Lost Ark</u>. That movie refers to the divine power linked to the Ark of the Covenant. After Solomon built the first temple at Jerusalem, they placed the Ark of the Covenant inside the Holy of Holies.

The End of the Original Sacrificial System

Perhaps you know of the word Babylon. They ruled the world from the late seventh century into the sixth century BC. During those times, the Jewish people chose to worship false gods. Moses had issued a prophetic warning of this event[2] many centuries earlier. According to the biblical texts, God destroyed the nation of Israel using the army of Babylon. That foreign army burned down Solomon's Temple in 586 BC. This event occurred on either Sunday, July 22, 586 BC, or on Sunday, August 19, 586 BC. We do not need to know the exact date of this event because the biblical texts reveal the 14,000 days sequence.

The following biblical texts help unwrap that period.

> [12]"Now on the <u>tenth day of the fifth month</u>, which was the nineteenth year of King Nebuchadnezzar, king of Babylon, Nebuzaradan the captain of the bodyguard, who was in the service of the king of Babylon, came to Jerusalem. [13]And <u>he burned the house of the Lord</u>, the king's house, and all the houses of Jerusalem; even every large house he burned with fire" (Jer. 52:12-13).

THE SCRIPTURE ABOVE SHOWS THAT SOLOMON'S TEMPLE BURNED down on the tenth day of the fifth Hebrew month. The end of Solomon's Temple marked the end of the sacrificial system.

For sure, the day of burning down Solomon's Temple gained the attention of the Jewish people. They began to observe that day as a fast. Even today, the Jewish people remember this day with fasting. They call it the Fast of Ab. Centuries later, Roman armies burned down Herod's Temple on this exact date in the year AD 70. Why would both Jewish temples be burned down on the Fast of Ab?

The answer is that we will find that the Fast of Ab is the final 14,000th day of the sacrificial system. In chapter twelve, we learned how to count 14,000 days quickly. The time is always thirty-eight years and four months.

The sacrificial system begins with the lamb selection on Nisan 10. Let's add four months to this day.

The first month, the tenth day (Select the sacrificial lamb)

1. The second month, the tenth day
2. The third month, the tenth day

3. The fourth month, the tenth day
4. The fifth month, the tenth day (the precise day Solomon's Temple burned down).

To get the 14,000 days, just count back thirty-eight years from the day Solomon's Temple burned down[3]. By doing this, we find the day on which people chose a sacrificial lamb. This sequence of 14,000 days repeats year after year. After the burning down of Solomon's Temple, Jewish people began an "official fast" in remembrance of that temple. The Fast of Ab may have begun as early as 585 BC.

This sequence of 14,000 days repeats year after year. On the 655th sequence in the year AD 70, Roman soldiers burned down Herod's Temple. As you read on, you will learn that Daniel's time-based texts about the future line up with the burning down of Solomon's Temple.

The Fast of Ab

Please understand that Solomon's Temple became the center point of the ancient Jewish faith. That building contained massive amounts of gold. At a higher level, the Jewish people used the temple for spiritual purposes.

The Jewish people are still keeping this fast. However, they do not appear to know of the 14,000 days related to that temple burning down.

There is much more to the Fast of Ab because prophetic texts foretell that the Fast of Ab is predictive of future events. A prophecy in the book of Zechariah foretells that the Fast of Ab and three other fasts will become "feasts of joy, gladness, and cheerfulness[4]." Why?

Could it be that the earthly temples were destroyed and replaced by an eternal temple that is not of this world?

Also, the Fast of Ab has yet to be called the Feast of Ab. Hence, the suggestion of future change on that very day.

What is Important to Remember?

The 14,000 days sequence for Solomon's Temple does happen on that same Ab 10 for the burning down of Herod's Temple. The 14,000 days period begins on the day that Jesus rides into Jerusalem, and people accept him as the Messiah. Consider that the time-based texts of Daniel conceal three things long before they happened.

- Messiah enters Jerusalem on Nisan 10, the day on which people select the sacrificial lamb.
- We find 14,000 days between the two foretold events.
- A foreign army burns down Herod's Temple on the Fast of Ab.

When this foretold cycle repeats, it will be the 655th recurring sequence.

15

SCIENTIFIC DATE FOR THE BEGINNING OF THE PROPHECY

When did the time-based prophecy begin?

Do we have credible source materials to find the precise date?

We have definite answers to both questions.

Firstly, we have legal documents from 2,500 years ago. At that time in Egypt, if you got married, sold land, or anything that could go to court, you needed a legal document. Hence, you hired a scribe to write a detailed paper, who would, on some occasions, use the Egyptian calendar as well as the Persian/Hebrew calendar.

These documents permit us to find the exact date that a scribe wrote them. Also, these documents refer to the king of Persia called out in biblical texts. These documents are called the Elephantine papyri.

Second, we have Babylonian astronomical texts from that era. Some of these refer to total lunar eclipses. We can know the exact hours of those events. Many of these documents refer to the Persian king in biblical texts.

These references allow us to find the exact date that King Artaxerxes issued the decree to rebuild Jerusalem. If you recall from chapter nine, this is the first event foretold in Daniel's time-based texts.

Finding the Date of the Decree to Rebuild Jerusalem

Here are the texts in Daniel about the start of the decree to rebuild Jerusalem. When that happened, we begin counting the days forward to future events. Each event must occur on a precise date. To begin, Daniel writes about the beginning of the prophecy that reveals when the Messiah will appear.

25"....you are to know and discern that from the issuing of a decree to restore and rebuild Jerusalem..." (Dan. 9:25).

When was this decree to restore and rebuild Jerusalem given[1]? We find the date in the following verse, which occurred in the springtime in the twentieth year of King Artaxerxes.

"[1]And it came about in <u>the month Nisan, in the twentieth year of King Artaxerxes,</u> . . .

[5]...I said to the king, "If it pleases the king, and if your servant has found favor before you, send me to Judah, to

the city of my fathers' tombs, that I may rebuild it" (Neh. 2:1, 5).

What do we need to date this event?

We need three things to date this event successfully.

- Find the twentieth year of Artaxerxes, the king.
- Obtain the date for Nisan 1 in that year.
- We know the day of the week to be Sunday[2].

When we use the exact science of astronomy to date all the ancient documents, we know when this king reigned over the Persian Empire. We also use total lunar eclipses found in Babylonian records[3].

Finally, we know the date on which someone assassinated Xerxes, the father of Artaxerxes, August 4, 465 BC. Here is the recorded text for that event.

"Month V, the 14th, Xerxes – his son killed him"[4]

If you want to dig into the details for the information above as well as all other documents, then you must read, <u>Beyond - The Coming Prince by Sir Robert Anderson</u>. That book is filled with details to obtain precise dates. I only give a shortened version within this book.

We know that the twentieth year for Artaxerxes began in March of 445 BC.

To give you insight into the scientific dating I use, consider the exact date from a document written only four months before this decree. The scribe wrote the text with double-dates on November 17, 446 BC. I have created a graph to explain this document.

Kislev 2 = Mesore 10 Year 19 of Artaxerxes I (Aramaic Papyrus No. 13)[5]

Kislev is the ninth Hebrew month, and it is the second day of the month.

The Egyptian calendar was very accurate. Every scholar will agree that the Egyptian date of Mesore 10 happened on November 17, 446 BC.

On the right side of the graph below, you can see the November 17, 446 BC date.

The Hebrew month and day, Kislev 2, means a new moon had occurred just before November 17. In the graph below, you will see only one day, which is the correct date[6].

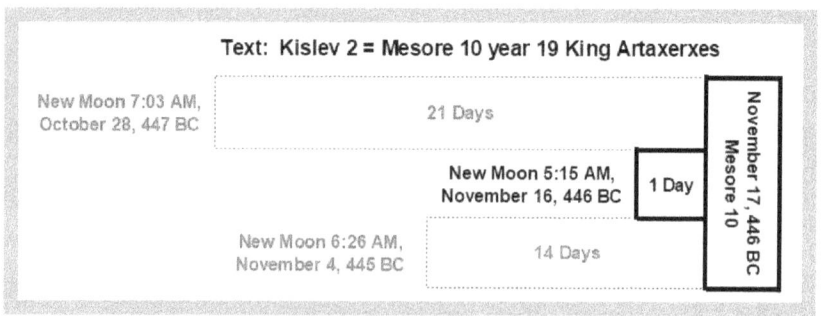

Scientific Date for Aramaic Papyrus No. 13

Based on this information, we must count forward to get an approximation for Nisan 1 (*the date referred to in Nehemiah 2:1*) based on the time of a new moon. The astronomical data reveals that the new moon for Nisan 1 in 445 BC occurred on Thursday, March 13, at 7:24 AM in Persia.

Two more information pieces are required.

Perhaps you know of the book entitled, <u>The Coming Prince</u>, written by Sir Robert Anderson. He thought that Nisan 1 occurred on Friday, March 14, 445 BC. I have concluded that this was the only error made by Sir Robert Anderson.

I base my research on the relationship between biblical texts, logic, and astronomical data. Daniel's time-based document supports that the decree must occur on Sunday, March 16, 445 BC.

In the next chapter, I will give proof that this date is the exact date for the decree.

16

BIBLICAL DATE FOR THE END OF THE SIXTY-NINTH WEEK

NISAN 10 - SELECTING THE SACRIFICIAL LAMB

Daniel's time-based texts allow us to find the date that Messiah would appear at Jerusalem. Based on the math, the time turns out to be Nisan 10. Perhaps you remember this is the day for selecting a sacrificial lamb at the first Passover in Egypt. This date becomes pivotal in starting the New Testament. Also, that the Old Covenant gets replaced.

If you think the Old Covenant is still good, please think about this question. Why were both Solomon's Temple and Herod's Temple burned down on the 14,000th day?

In chapter fourteen, we learned about the importance of the dates, Nisan 10 and Ab 10. Both times repeat in the time-based texts for the future. We discover that the Old Testament sacrificial system happens on these precise days surrounding the life of Jesus.

The prophecy foretells of a suffering Messiah because the texts refer to atonement. Jesus' death on the cross on The Passover equates to the foretold atonement.

This means the time-based texts, grounded on scientific dating, foretell details about the New Covenant[1]. We begin to understand the purpose of the time-based passages.

There is much more to this than just repeating the exact dates of earlier events. At a higher level, the foretold events reveal spiritual knowledge. In this case, the burning down of Herod's Temple on the 14,000th day, foretold many centuries before that event happened.

The Time-Based Calculation

Below are the passages that unveil the time for the appearance of Jesus on Palm Sunday (Nisan 10) as the Lamb of God.

"….you are to know and discern that from the issuing of a decree to restore and rebuild Jerusalem (Date: Sunday, March 16, 445 BC) until Messiah the Prince there will be seven weeks and sixty-two weeks (Calculated Date: Sunday, April 6, AD 32)" (Dan. 9:25).

I will repeat Sir Robert Anderson's calculations found in The Coming Prince. Since I want to keep this book an easy read, we only need to add 173,880 days to the beginning date for the time-based texts. In chapter fifteen, we show this began on Sunday, March 16, 445 BC. Should you want to understand the biblical math from the time-based texts, I have placed the math into this footnote[2].

We learn that Jesus entered Jerusalem on Palm Sunday. The date was Sunday, April 6, AD 32[3]. Please understand that we arrived at this date based on the math in the time-based texts. Also, we used the biblical constant of 360 days in a year. As stated in chapter eleven, this biblical number isolates precisely 14,000 days in the time-based texts.

I am intrigued by what Jesus did on this day. Without knowing it, the New Testament author wrote of Jesus' foretelling of the coming destruction of Herod's Temple exactly 14,000 days from that very day.

> [41]"...when he approached, He saw the city and wept over it, [42]saying, "If you had known in this day, even you, the things which make for peace! But now they have been hidden from your eyes. [43]"For the days shall come upon you when your enemies will throw up a bank before you, and surround you, and hem you in on every side, [44]and will level you to the ground and your children within you, and they will not leave in you one stone upon another, because you did not recognize the time of your visitation" (Luke 19:41-44).

Jesus' words foretell of the coming destruction of Jerusalem. His words align with Daniel's "time-oriented" prophecy.

> "...the people of the prince who is to come will destroy the city (*Jerusalem*) and the sanctuary (*the temple*)" (Dan. 9:26).

Verifying the Decree to Restore Jerusalem on Sunday, March 16, 445 BC

The website with the "Julian Day and Civil Date Calculator" allows us to confirm the decree to rebuild Jerusalem happened on a Sunday. On the left side of the table below, the date of Friday, March 14, 445 BC, is used. Sir Robert Anderson selected the Friday date. On the right side, I placed the calculated Sunday date. Please take note that the second row validates 173,880 days lines up with Sunday, March 16, 445 BC. In turn, this will result in precisely 14,000 days to the burning down of Herod's Temple.

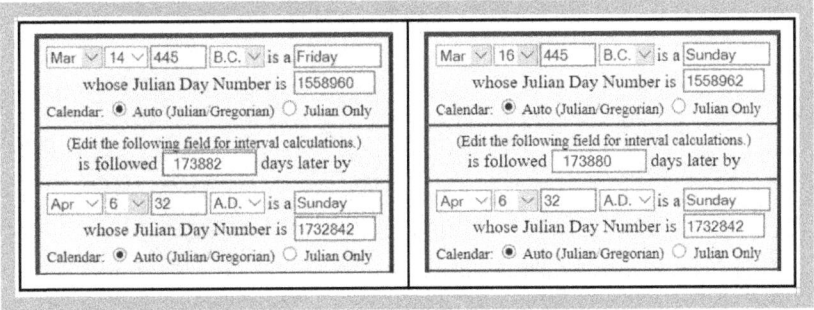

~

Validating AD 32 – The Evidence

I base my research on The Coming Prince by Sir Robert Anderson. The Palm Sunday date of April 6, AD 32, is in that book. Most scholars have concluded that Jesus' crucifixion could only happen in either AD 30 or 33.

Scholars reject AD 32 based on assumptions about astronomy. Scholars think the crucifixion must occur in a year that the full moon happens on a Thursday or Friday. For the year AD 32, this assumed requirement is out of place.

To counter this issue, chapter eighteen shows that AD 32 is the year of the crucifixion. I have found information with supporting data that reveals how inconsistent the Hebrew calendar was before the destruction of Herod's Temple. The variation shows that the year AD 32 is very credible. In fact, Daniel's time-based texts make the Bible relevant across history. This is especially true as the passages were recorded centuries before the predicted events came to pass.

17

SCIENTIFIC DATE FOR BURNING DOWN HEROD'S TEMPLE

AB 10 - DESTRUCTION OF EARTHLY TEMPLES

Daniel's time-based texts foretell the destruction of the temple in Jerusalem. The sequence shows the temple burning down "after the Messiah" had appeared at Jerusalem. Consider the passages.

> "...Then after the sixty-two weeks the Messiah will be cut off*(the crucifixion event)* and have nothing, and the people of the prince who is to come will destroy the city (*Jerusalem*) and the sanctuary (*the temple*)" (Dan. 9:26).

For the date of burning down Herod's Temple, two ancient authors wrote of it. They wrote details that allow us to find the exact time for burning down Herod's Temple. To begin, Josephus was an eyewitness. He wrote in a way that strongly suggests the event occurred on a Sunday.

About AD 150, Rabbi Yose wrote that both Solomon's Temple and Herod's Temple burned down on a Sunday[1].

We also know that it was the tenth day of the fifth Hebrew month[2]. We only need to align ten days from a new moon to a Sunday in August, AD 70.

In AD 70, a new moon happened on July 26, about 3:31 AM in Jerusalem. We find this aligns ten days to Sunday, August 5, AD 70. The evidence and logic permit other people to arrive at this same date. Is it any wonder that Dr. Jack Finegan used this evidence, logic, and astronomy to arrive at the exact date I calculated in the 1980s[3].

The previous chapter shows that Jesus entered Jerusalem as the Messiah on Sunday, April 6, AD 32. We find precisely 14,000 days from that Palm Sunday day to the burning down of Herod's Temple. The following graphic of the Julian calendar verifies the 14,000 days as follows.

```
Apr  6  32        A.D.  is a Sunday
     whose Julian Day Number is 1732842
Calendar: ● Auto (Julian/Gregorian)  ○ Julian Only

(Edit the following field for interval calculations.)
    is followed  14000   days later by

Aug  5  70        A.D.  is a Sunday
     whose Julian Day Number is 1746842
Calendar: ● Auto (Julian/Gregorian)  ○ Julian Only
```

You may also find it thought-provoking that Titus, the Roman general, ordered his soldiers to put out the fire that engulfed the temple. However, the soldiers did not obey Titus since they despised the Jewish people and lusted for the gold that adorned the holy building. [Wars of the Jews, 6.4.7].

In summary, Josephus wrote about the beauty and vast wealth of the temple. Shocked by the temple's destruction, Josephus wrote,

> *"one might* comfort himself with this thought, that it was fate that decreed it to be, which is inevitable, both as to living creatures and about works and places also. However, one cannot but wonder at the accuracy of this period thereto relating; for the same month and day were now observed, as I said before, wherein the holy house was burned formerly by the Babylonians." [Wars of the Jews, 6.4.8].

God-like Control of Humans

The time-based texts do not give us math to calculate the number of days from Palm Sunday to the burning down of Herod's Temple. Despite this issue, there are precisely 14,000 days between the two events. On top of this idea, the dates are repeated for the burning down of Solomon's Temple.

To achieve the 14,000 days requires divine control of some humans. It appears that God swayed some people to cause these events to occur on these exact dates. Why?

Did God orchestrate both temple destruction events to occur on the final 14,000th day?

In my opinion, that is what happened.

If you review chapter thirteen, you learn that atonement conveys the purpose of the time-based texts. It appears that Jesus' sacrificial death and resurrection replaced the earthly temples.

Spiritual Source of the Qur'an

Consider the source of the Qur'an as this research supports Jesus' death and resurrection by foreign armies burning down both Solomon's Temple and Herod's Temple on the 14,000th day. This information reveals that the source of Islam does not come from a transcendent Being based on two ideas.

1. The Qur'an has no texts that foretell of specific future events that will happen on an exact date, centuries into the future. Hence, no evidence that Allah transcends time.
2. The final 14,000th day incident reveals spiritual truth.
3. Why does the Qur'an teach the diametrically opposite view of Jesus?

Moreover, the Qur'an teaches that Jesus is not the Son of God. In contrast, consider the time-based texts of the Messiah with other Old Testament ideas. For example, you can study the Messiah as the Son of God in Psalms 2:7. I suggest that you explore all of Psalms seven as well as Proverbs 30:4. These biblical texts recorded long before Jesus, validate the New Testament view of Jesus.

Daniel's Time-Based Prophecy
Why Sunday Dates?

Perhaps you will find it thought-provoking that every foretold date occurs on a Sunday, the first day of the week. The Old Covenant foundation is the number seven and the Sabbath, which always occurs on Saturday. However, in the New Covenant, Jesus rises from the dead on Sunday. Is it possible that Daniel's "time-oriented" prophecy is prophetic of the New Covenant[4]?

Moses also wrote about the Feast of First Fruits only occurring on Sunday.

> [10]"When you enter the land which I am going to give to you and reap its harvest, then you shall bring in the sheaf of the first fruits of your harvest to the priest. [11]And he shall wave the sheaf before the Lord for you to be accepted on the day after the Sabbath (Sunday) the priest shall wave it" (Lev. 23:10-11).

The Apostle Paul knew about the Feast of First Fruits. He wrote about the First Fruits with a direct link to the resurrection of Jesus on the first day of the week.

> "…Christ has been raised from the dead, the first fruits of those who are asleep" (1 Cor. 15:20).

18

EVIDENCE THAT VALIDATES AD 32

Scholars instantly reject AD 32 as the year Jesus died on the cross. Because they believe this could only occur in a year with the full moon occurring on a Thursday or Friday. In the year AD 32, the full moon happened on Monday. However, I have found information that shows AD 32 is the year of the crucifixion.

Scholars have made false assumptions. I discovered they do not use all data for the sun, the moon, and the earth. If you make decisions without using all the data, you will make poor decisions. Are you ready to learn why the scholars are wrong?

I want to keep this book an easy read. If you're going to delve into all the data, please read book three in the Expect to Live Forever book series, Beyond - The Coming Prince by Sir Robert Anderson.

Mishnah Texts on Variation of the Persian/Hebrew Calendars

SCHOLARS HAVE CHOSEN TO IGNORE SPECIFIC EVIDENCE THAT refutes their views. If they were correct, then people living at the time of Jesus were very advanced on astronomy. However, that is a false belief. In truth, those ancient people did not understand astronomy that well. Advances in the Hebrew calendar took place almost 400 years after Jesus lived.

Before the Romans burned down Herod's Temple, the Pharisees talked about their laws. They called this the oral law. They used the oral law as a way of knowing when to observe the Passover. However, after the burning down of Herod's Temple, they wrote the verbal code on paper. They published this collection of laws about AD 200. They called it the Mishnah.

In the Mishnah, we learn how they observed the Passover and other feasts. The Mishnah tells us how well the Pharisees understood astronomy. We learn that they did not know it that well. The information in the Mishnah makes AD 32 a viable year for Jesus' crucifixion.

Who would you trust more?

People who knew their environment and wrote about it. Or do you trust a person living 2000 years later that decides they can reject what those people wrote, mainly based on modern-day knowledge?

At the end of the nineteenth century, scholars did some work on calendar variation. However, it appears they concluded that the deviations must not be valid. I often read scholars that claim the people from long ago were writing theories! It seems that scholars want to maintain their views of ancient history.

In contrast, I use the positions of the sun, the moon, and the earth to show that they did write down what happened. The Pharisees did not use astronomy at the high-level that scholars have assumed.

Perhaps the most significant difference is the source of the data. I use all the data, including recently published materials printed in the twenty-first century. In contrast, scholars created their paradigms in the nineteenth and twentieth centuries. Therefore, I address these weak paradigms. My tools are logic based on the positions of the sun, the moon, and the earth.

False Paradigm No. 1: Crucifixion with a Full Moon on a Thursday or Friday

The Mishnah shows how poorly the Pharisees observed their calendar. Consider this quote from the Mishnah as follows:

> "They do not count less than four full months in the year and [to sages] have never appeared more than eight" [Neusner, Jacob, Mishnah, Arakin 2:2a].

In the text above, "four full months" reveals four months with 30 days each. If the month is not "full," then that month will be twenty-nine days in length.

A year with four full months will also contain eight months of twenty-nine days. When a year has eight months with twenty-nine days, the calendar will be off by two entire days.

Back in those days, the year could range in length from 352 to 356 days (a five-day range). In contrast, the modern Hebrew calendar ranges from 353 to 355 days (a three-day range). So, when Jesus walked the earth, the Hebrew calendar had sixty-seven percent more variation.

We will learn that this makes the year AD 32 a viable crucifixion year. Is there any data to support this viewpoint?

I found data that verifies the statement in the Mishnah is accurate. In the years 226 through 221 BC, I found a sequence of sixty-two months with the number of days in each month. Scholars translated this data from Babylonian Cuneiform tablets[1]. In the years 224 and 223 BC[2], I discovered three series where there are eight months of twenty-nine days and four months of thirty days[3]. If you want to review this data, I have placed it in book three of the Expect to Live Forever book series, <u>Beyond - The Coming Prince by Sir Robert Anderson</u>[4]. This evidence verifies that the statement in the Mishnah is true and accurate. Let me be emphatic, the report in the Mishnah is a fact, not a theory.

In AD 32, the calendar was out of line by two negative days. That means that the people did not know the exact time of the new moon. Perhaps more to the point, they did not care to be that accurate. I also found evidence to validate this point.

The negative days happened more often than scholars would believe. I will show actual misalignments later on. It is certain that nowadays, scholars have weak assumptions. I often read that they blame the scribes for human error. In truth, scholars' should look in the mirror before they blame people of the ancient world.

Data proves the Mishnah statement is true. We will learn that this confronts some things that scholars teach about chronology. To be specific, the idea that the crucifixion must occur in a year where the full moon must occur on a Thursday or a Friday is not valid. Scholars have created a false paradigm. This paradigm makes sense if you ignore all the information and data.

False Paradigm No. 2: No Difference between Persian Area to the Egyptian Area

If you traveled from Persia to Egypt, it took five months. Communication across the vast expanse of the Persian empire was poor. The calendars in Egypt could be an entire month different from the Persian area.

Some scholars[5] agree. However, that scholar blames this difference on the scribe. In contrast, my analysis supports that the ancient authors rarely made errors.

Instead of blaming ancient scribes, accept their data. Then compare differences between the two areas. I have compared the two sets of data. We have large amounts of Persian area data to compare to twenty-two items written at Elephantine, Egypt. Analysis of this data shows poor coordination, mainly during the initial twenty-five years of King Artaxerxes (465 to 440 BC).

To support this problem, I found years with large amounts of calendar variation. The range is high for the years 250 to 217 BC. When we use that data for a guideline, we begin to understand why scholars are limiting our knowledge[6].

For example, scholars use a reference book[7] with minimal variation. The low range in data limits a scholar's ability to understand when events happened in ancient history. In my estimation, the reference book used by scholars is off up to about 97 percent[8].

False Paradigm No. 3:
The Practice of the Biblical Feasts

How important was it that the Jewish people observed the Passover on a perfect date relative to the full moon?

In the Mishnah, we learn how the Pharisees viewed this question. Consider this quote from the Mishnah, which refers to the Old Testament to support its viewpoint.

> "... I can provide grounds for showing that everything that Rabban Gamaliel has done is validly done, since it says, *"These are the set feasts of the Lord, even holy convocations, which you shall proclaim"* (Lev. 23:4). Whether they are done in their proper time or not in their proper time, I have no set feasts but these [which you shall proclaim]." [Neusner, Jacob, Mishnah, Rosh Hashanah 2:9a].

From the Mishnah, we learn that the exact day was not that important. The important thing was to "proclaim a day for a feast" so that everyone would know to observe it on the day the authorities declared it to happen. Therefore, the feast days did not have to be exact. We begin to grasp the false paradigm of requiring The Passover to occur on a Friday with a full moon.

There seems to have been some leeway for error in how people observed the biblical feasts. Hence, another idea that adds support for Jesus' crucifixion in AD 32.

What must happen for the date of The Passover?

The only requirement for the Passover was green barley heads. People collected and roasted the barley for the Passover. Green barley means that the grain heads had developed, but they were still green.

Nowadays, some people claim the Passover must always happen after the Springtime equinox. However, data from ancient history shows this is false. I have found data that validates the Passover did get observed before the Springtime equinox.

What if the grain ripened early?

If this occurred, the time for Passover would happen earlier in the springtime. In a farming society, people would notice the grain ripening early.

What if the grain ripened late in the spring due to colder weather?

When it was cold, and the grain did not begin to create green barley heads, then the Passover would be moved forward a month. They added an extra month, which is called an intercalary month[9].

Based on the data from the third century BC, the Passover would have occurred as early as March 12 in 221 BC or as late as May 20 in 246 BC. This range is almost two and a half months across twenty-five years.

How much influence did the weather have on the timing of the Passover?

Nowadays, global warming is becoming a political issue. A graph of global warming from 2500 BC to nowadays shows how weather changes affected people in ancient history.

I suggest you review the temperature changes from 250 to 200 BC in this graph. Please note that it was getting warmer during the time for which we have calendar data[10]. I find it thought-provoking that the late Passovers from 250 to 228 BC were in the colder phase of the graph. Then from 228 to 217 BC, the dates for Nisan 1 shifted to earlier in the year. The Passover took place before the Spring equinox. This shift in the calendar dates supports that global temperatures did increase at that time. Temperature affects the growing season of barley used during the Passover.

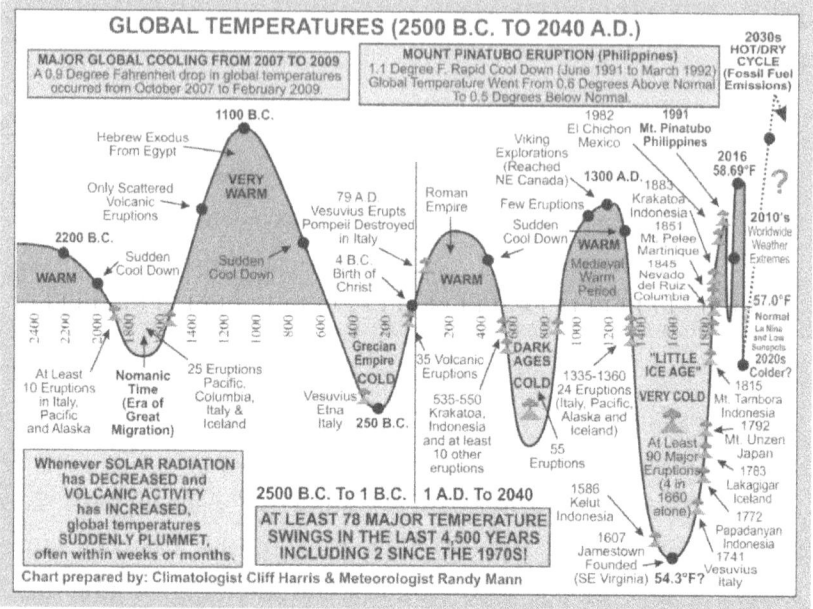

Global Warming and Cooling Influence Observance of the Passover in Ancient History

The best example that Jesus died on the cross in AD 32 comes from the year 459 BC.

To begin, here is an Elephantine papyrus that scholars could not date correctly. The reason they failed is due to their weak assumptions.

In the following quote, Kislev is the ninth Hebrew month. Mesore is the twelfth month of the Egyptian calendar. The numbers after each month are the day of that specific month. I will explain this text.

Kislev 21 = Mesore 1, year 6 of Artaxerxes I (Aramaic Papyrus No. 8)[11]

Eternal Life - why you should expect to live forever

Consider what these scholars wrote.

"The papyrus is well preserved and creates no reading problems. However, the dates as given can be made to agree by no known methods, so that a scribal error must be involved[12]."

Why did the two scholars blame the scribe?

Please understand that the Egyptian calendar was very dependable. Every scholar will agree with the date. In contrast, scholars will disagree about the Hebrew calendar with its variation. We know the exact time for the daylight portion on which the scribe wrote the legal document. The Egyptian month is called Mesore. It is the first day of the month, which means the scribe wrote this document during the daylight hours of November 11, 460 BC.

The document refers to the month named Kislev, and it's the twenty-first day. We expect to find a new moon twenty-one days before November 11. The new moon did occur on October 21, 460 BC, at 2:12 AM at Elephantine, Egypt. To align the date to November 11 requires that Kislev 1 begins at sundown on October 20.

Below is a graphic that will help you understand this dating at a simple level. Notice that the middle row contains twenty-one days from the new moon to the November 11 date for the year 460 BC. The years before and after the year 460 BC do not have the twenty-one days alignment. When sequences of these are aligned, the exact times are credible. This alignment evidence strongly supports the data from the third century BC[13] give us a reliable guideline. The position of the sun, the moon, and the earth verify the scribe wrote accurately.

In contrast, the scholars rejected this date based on weak presuppositions.

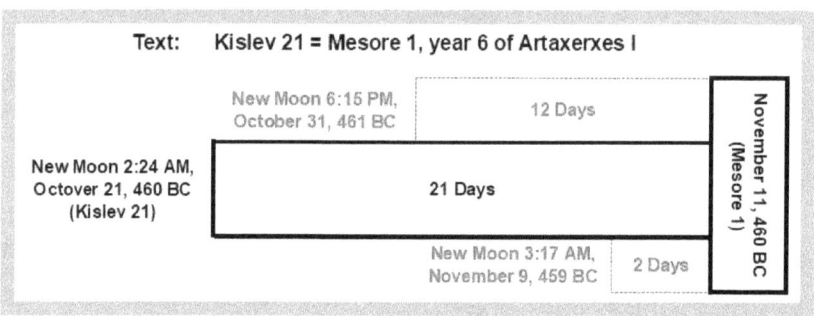

Scientific Date for Aramaic Papyrus No. 8

The guideline shows the date of October 20 is acceptable. The earliest possible time for Kislev 1 would be October 19. Although October 20 is early, it meets the criteria.

Now we can begin to understand why the scholars have rejected this date as a scribal error. In the minds of these scholars, it is too early in the year for the month Kislev to occur. It is essential to point out that the twenty-one day new moon alignment provides scientific support for the November 11, 460 BC date.

The second problem is the fact that there is a negative translation time of eight hours and twelve minutes. The Mishnah's statements make unfavorable translations acceptable. We expect to observe negative translation times. Therefore, the date that the scribe wrote down is credible.

Based on the Mishnah, this supports that the year 460 to 459 BC would be a year with eight months of twenty-nine days and four months of thirty days.

Now consider that there were two documents written by the scribe during this month. The scholars rejected it for the same reasons, which is more evidence of a false paradigm used by scholars.

Another issue is that scholars have not realized the vast area of the Persian empire caused poor communication to the far reaches of the empire. I have added a map showing the travel route from Persia to Egypt. We know from the books of Ezra and Nehemiah that it took four entire months to travel from Persia to Jerusalem. I have created a graphic to illustrate this idea. Travel to Elephantine, Egypt took another month beyond Jerusalem.

Based on poor communication, the calendars at various locations did have different dates than those in Persia. This issue seems to have been accepted by the Persians. However, by 440 BC, this communication problem stopped. I have found scholarly support for this difference[14].

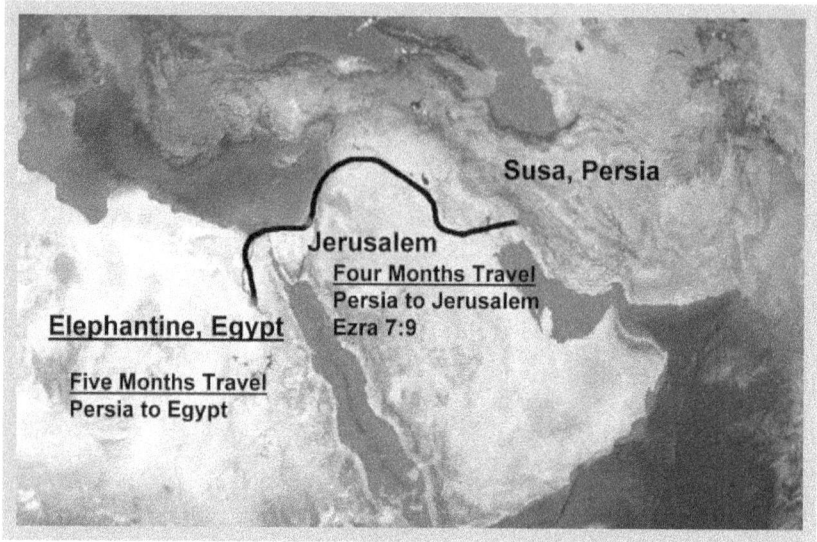

*Five Month Travel Route from Susa, Persia to Elephantine, Egypt
(Fifth Century BC)*

Since we have found what the scribe wrote as being acceptable, we will now begin to realize that the negative value for translation is relevant to the year AD 32 as the year of Jesus' crucifixion. Because the next item is the Cairo Sandstone Stele with the double-dated time references on a public monument. In contrast with a legal document that a few people would read, lots of people would read the text on display for the public. Therefore, the scribe would make sure there were no errors.

Consider the text written on the Cairo Sandstone Stele. In this text, Sivan is the name of the Hebrew month, and Mechir is the name of the Egyptian month.

Sivan = Mechir, Year 7 of Artaxerxes I[15]

Consider how the scholars discard this valuable reference.

"Because of the wide range of this date and its ambiguity, this stele does not settle the problem raised by Aramaic Papyrus No. 8. If the 7th year of Artaxerxes is recorded here according to the Egyptian system of reckoning, as is most likely the case..." [16]

I stopped at the scholars referencing the Egyptian calendar because that is correct. The Cairo Sandstone Stele is strictly Egyptian. It aligns perfectly with the papyrus we dated to November 11, 460 BC.

The grouping of the Cairo Sandstone Stele to the November 11, 460 BC papyrus requires an intercalary month in the spring of 459 BC. There were no scribal errors.

Below is a graphic that will help you understand the scientific dating for the Cairo Sandstone Stele. Notice that the middle row contains a negative time close to two entire days[17].

In the graphic below, the years before (460 BC – 18 days) and after (458 BC – 10 days) do not have an agreeable alignment for the new moon to May 15. The precise date presented below is credible.

Scientific Date for the Cairo Sandstone Stele

Why did the scholars reject the Cairo Sandstone Stele?

There are three issues to explain WHY they could not resolve the precise date, which is similar to the earlier discussion.

First of all, it turns out that the month of Sivan is too early in the year for the scholar's viewpoint to accept. In contrast, the guideline data from the third century BC data shows this is an acceptable date. Sivan 1 could be as early as April 26. Since it is May 14 for the Hebrew calendar to begin at sunset, the Hebrew date on the Stele is acceptable. Weak assumptions blind the two scholars. They move the year on the Stele forward into 458 BC, then end up rejecting it outright. Remember this is a public monument and it has to be correct!

The Egyptian date is Mechir 1, which means it was the daylight hours on May 15, 459 BC.

The primary issue is the fact that the chronology of these two scholars is off by an entire year[18]. They could not make their weak assumptions align to the year 459 BC.

The new moon occurred on May 16 at 5:05 in the afternoon near modern-day Cairo, Egypt. For the date of Sivan 1 to align to May 15 means that the day began at sunset on May 14.

Therefore, Sivan 1 has a negative translation of one day, twenty-three hours, and five minutes[19]. This unfavorable translation would be "unthinkable" for these scholars, which is the third issue.

Astronomical Support for the Crucifixion in AD 32

How great was the unfavorable translation in the year AD 32?

For sure, the year AD 32 requires eight months of twenty-nine days. The actual time is a negative 2.18 days alignment. The time is in-family with the Cairo Sandstone Stele negative value of 1.96 days.

We do not have any astronomical data to validate the year AD 32. However, we have proven that a full moon happening on a Monday is acceptable for the year of Jesus' crucifixion. The factors in support of this viewpoint are based on astronomical data and evidence from the Mishnah.

Firstly, the key issues are that it was not that important to have some divinely ordained perfect day for the Passover feast. For example, if the Passover was off two days, would it matter? According to the Mishnah, this is acceptable.

Second, the Hebrew calendar was highly variable and did have years where there were eight months of twenty-nine days. I found data to validate this. Also, the Cairo Sandstone Stele corroborates that this did happen.

Moreover, I found astronomical data from the third century BC that validates these concepts.

Negative translation times continued up to 300 years after the

Romans burned down Herod's Temple in AD 70. Jewish people began to use the calculated Hebrew calendar about AD 360.

The calculated Hebrew calendar did away with years that would have eight months of twenty-nine days. For example, the time from the Passover to the seventh month became fixed.

Below is a graphic that will help you understand the year of Jesus' crucifixion based on astronomy. I have included the years from AD 30 to 33. These are the most familiar years that people think Jesus died on the cross.

The year AD 32 stands out with an unfavorable two-day translation. This aligns 14,000-days to Romans burning down Herod's Temple (shown in the far-right column).

Crucifixion Astronomy

Time for the New Moon (Jerusalem Locale)	Saturday SunSet Julian Calendar Date (begins Nisan 10)	Palm Sunday Julian Calendar Date Dayfight Hours	Translation Time (Days) from New Moon to Nisan 1		Days to Destruction of Second Temple (Nisan 10 to Av 10)
New Moon 8:08 PM, March 22, AD 30	April 1, AD 30	April 2, AD 30	0.9111	Selecting the Lamb (Palm Sunday) Nisan 10	14735
New Moon 1:54 PM, April 10, AD 31	April 21, AD 31	April 22, AD 31	2.2688		14350
New Moon 10:22 PM, March 29, AD 32	April 5, AD 32	April 6, AD 32	-2.1819		14000
New Moon 1:00 PM, March 19, AD 33	March 28, AD 33	March 29, AD 33	0.2083		13643

Based on this research, two factors stand out regarding getting the negative 2.18 days for the year AD 32.

1. The calendars always had the potential of having eight months of twenty-nine days, which would cause a potential two-day negative translation for the new moon.
2. The Pharisees did not have a problem with the Passover occurring precisely on time with a full moon. They

could shift the day regarding the day of the week they chose to have this assembly of people take place. This factor adds to the unfavorable position of the moon for the Passover in AD 32.

Overview of Important Points

Here is a very long bullet list that supports the year AD 32 for Jesus' sacrificial offering on the cross. The reason it is quite long is that many factors validate the year AD 32. Also, I am sure that the scholars have a paradigm based on their hearts, as opposed to their minds.

My approach is data-based as opposed to paradigms based on opinion without supporting astronomical data. Also, the time-based texts predicted specific events to occur on exact dates, centuries into the future.

1: Mishnah's statements reveal that the Hebrew calendar at times had years that contained eight months of twenty-nine days with only four months of thirty days' length. I documented four unfavorable translations of two whole days, which includes the Cairo Sandstone Stele for the year 459 BC, as well as Babylonian Cuneiform texts[20].

2: Babylonian astronomical texts from the third century BC verify that the statements in the Mishnah are accurate and correct. I show how this data validated the Aramaic Papyrus #8 dated November 11, 460 BC. Scholars have erroneously chosen to blame the scribe with human error.

3: We know the beginning and end of the original sacrificial system. Moses' instruction to select a lamb on the tenth day of Nisan began this system. Then it ended on the day a foreign army burned down Solomon's Temple in 586 BC on the tenth day of Ab. As a result, we discover the 14,000-days alignment that has now repeated itself for more than 2,604 years. Year after year, this alignment has occurred since the destruction of Solomon's Temple. Each year, the Jewish people remind themselves by observing the Fast of Ab. The 655th cycle began with Jesus' entry into Jerusalem as Messiah on Palm Sunday and ended on the day that Roman armies burned down Herod's Temple.

4: The details of temple destruction events strongly support that these were acts of God. There are common factors for these independent events. Below is the list of these factors from Rabbi Yose, the student of Rabbi Akiva:

- On the same Hebrew calendar date, the tenth day of Ab
- On the same day of the week, Sunday
- After a Sabbatical year
- During the priestly division of Jehoiarib

5: The fact that we can isolate a mystical period of 14,000 days hidden deep in the biblical texts at the time of Moses and Joshua, which supports God-breathed those texts.

6: The Jewish people crossed into the Promised Land on the 14,000th mystical day at the time of Moses and Joshua.

7: The fact that we can isolate an actual period of 14,000 days hidden in the texts of Daniel's "time-oriented" prophecy to the destruction of Herod's Temple. Support for this knowledge comes from the research of Sir Robert Anderson[21] and the precise date for burning down Herod's Temple done by Dr. Jack Finegan[22].

8: Roman soldiers destroyed Herod's Temple on the 14,000th day, which aligns with Nisan 10, the date of Palm Sunday when Jesus publicly acknowledged himself as the Messiah.

9: The fulfillment of Jesus' prophecy about the Jewish people aligns 14,000 days to the Feast of Trumpets in the 21st century. NOTE: From June 6-7, 1967 to October 4, 2005. The final date includes an alignment to a solar eclipse. We have observed a 14,000-day arrangement to the Feast of Trumpets unveiled in the modern era. Chapter nineteen presents this information.

10. Eight copies of the book of Daniel with the oldest dated to 125 BC supports that the texts are from an eternal transcendent Being that knows future events.

11. Use of the biblical constant, 360 days per year, unveils the 14,000-day sequences at the time of Moses and Joshua as well as in Daniel's "time-oriented" prophecy. We have repeatability for the use of 360 days per year, which always results in finding a 14,000 days sequence.

12. That sabbatical years in Daniel's "time-oriented" prophecy link to the period of 14,000 days.

13. That 14,000-day intervals have been sealed away from Bible scholars for millennia.

14. That the original sacrificial system, from start to end, is a mystical period for repeating 14,000-day sequences that occur over and over on exact dates of the Hebrew calendar:

- Old Covenant (Nisan 10 to Ab 10, with 38-year intervals, link to the Fast of Ab)
- New Covenant (Nisan 10 to Ab 10 link directly to

> Daniel's "time-oriented" prophecy ending on the Fast of Ab)

15. Old Testament 14,000-day lengths are both mystical and real-world in kind.

16. New Testament 14,000-day lengths align only to real-world historical events.

The order of scripture documented herein supports that God orchestrated temple destruction events. This information points humanity to an eternal temple that is not of this world.

19

JESUS' SECOND COMING PROMISE

During the Passover week at which Jesus died on the cross, he foretold of the coming destruction of Herod's Temple. The ideal text for this prophecy is the gospel of Luke. It begins with the disciples of Jesus bragging about the beauty of Herod's Temple.

> 5"… while some were talking about the temple, that it was adorned with beautiful stones and votive gifts, He said, 6'as for these things which you are looking at, the days will come in which there will not be left one stone upon another which will not be torn down.' 7And they questioned Him, saying, 'Teacher, when therefore will these things be? And what will be the sign when these things are about to take place?" (Luke 21:5-7).

Jesus answers their question with a series of signs that will occur in the future. However, the seventh verse reveals the most important thing about these texts. The disciples asked for a single sign. That sign, which occurs only one-time, will contrast with a plethora of events that keep repeating.

What are the signs that keep repeating?

These include wars, earthquakes, plagues, famines, and being persecuted for your faith in Christ. These are signs that repeat many times. These signs cannot be isolated to a specific era because they are always happening.

However, there is a set of verses that reveal a single sign. The foretold event will only happen one time. It answers the disciples' question, *"What will be the sign?"*

Before going further, you need to understand that the disciples were not asking Jesus when he would return to earth in a second coming. Luke, the author, investigated before he wrote[1]. Perhaps there were long discussions about what happened on that day as Jesus talked. The way Luke framed the question is very different from how the gospels of Mark and Matthew frame the disciple's question. The answer to Luke's question has already happened. In contrast, the answer to the question in Mark and Matthew will occur in the future, at an unknown time.

A paraphrase of the text in Luke would state, *"What will be the sign that Herod's Temple is about to be destroyed?"*

We find the answer to the disciples' question in verse twenty.

"...when you see Jerusalem surrounded by armies, then recognize that her desolation is at hand" (Luke 21:20).

The single sign of the future destruction of the temple would occur when Roman armies surrounded the city. This event was temporary. A Roman army surrounded the city, then withdrew. When the Roman army withdrew, the Christ-followers fled Jerusalem. Then the Roman armies returned around the Passover in the Springtime of AD 70. This time, they encircled the city with its inhabitants. There was no escape. The Roman army destroyed Jerusalem with enormous bloodshed.

Then the texts in the gospel of Luke take us well beyond AD 70. Luke indeed writes details that bring the question into the twenty-first century. We learn in verses twenty through twenty-four that Jesus links the destruction of Herod's Temple in AD 70 to future events. Take note that Jesus reveals the future dependent on the whereabouts of the Jewish people.

Jesus' words foretell of the dispersion of the Jewish people around the world.

> "... they will fall by the edge of the sword, and will be led captive into all the nations..." (Luke 21:24a).

This dispersion began in AD 70. How long would this last?

Jesus' words clearly state that the Jewish people will return to their homeland and regain the political and military control of Jerusalem. Hence, we discern the key is whether Jewish people or Gentile people control Jerusalem. Gentiles began to control Jerusalem when Rome destroyed the city in AD 70. The end of Gentiles controlling Jerusalem happened when modern-day

Jordan lost control of Jerusalem in the Six-Day War on June 6-7, 1967.

> "... they (*the Jewish people*) will fall by the edge of the sword, and will be led captive into all the nations; and Jerusalem will be trampled underfoot by the Gentiles until the times of the Gentiles be fulfilled" (Luke 21:24).

What happened in the Six-day war turned out to be the fulfillment of Jesus' prophecy about the Jewish people. Because Gentiles lost control of Jerusalem[2]. What does this mean for us?

Just add 14,000 days to the June 6-7 date, and you will end on October 4, 2005. This date was Rosh Hashanah, the Feast of Trumpets, which symbolizes the return of Jesus to earth. Both the event of Jewish people gaining Jerusalem with the perfect alignment to the Feast of Trumpets answers the question in Luke 21:7, *"What will be the sign?"*

At this point, please consider that the question is about two eras that show God-like control of human events. Because we find there are 14,000 days for the following two periods:

- From Palm Sunday to the burning down of Herod's Temple (from Sunday, Apr. 6, AD 32 to Sunday, Aug. 5, AD 70).
- From the Six-Day War retaking of Jerusalem to the Feast of Trumpets (from Tuesday, June. 6/7, 1967 to Tuesday, October 4, 2005).

I created a graph to help us understand the repeating 14,000

Eternal Life - why you should expect to live forever

days in the gospel of Luke. On the left side of the black area is the generation that witnessed the destruction of Herod's Temple.

On the right side of the black area is the fulfillment of Luke's gospel[3] in the modern era. The black area between these two times marks "Gentile Control of Jerusalem." That time was unknown until the Six-Day War event that fulfilled Jesus' prophetic words to the letter.

Nisan 10, Palm Sunday, April 6, AD 32 End of Daniel's 69th Week Jesus enters Jerusalem as Messiah	14,000 Days	Ab 10 - Fast of Ab, Sunday, August 5, AD 70 Romans burn down Herod's Temple	Jesus' Prophecy and Daniel's Prophecy coincide Luke 21:7, 20-24 and Dan. 9:26	Luke 21:24b (Unknown Time Period) Jewish People scattered globally Gentiles control Jerusalem for 1,897 Years. (692,806 Days)	Jewish People Regain Control of Jerusalem June 6/7, 1967 Six-Day War Jesus' Prophecy fulfilled Luke 21:7, 24b Sign that Second Coming is "about to take place"	14,000 Days	Feast of Trumpets, Tuesday, October 4, 2005
Ancient History Fulfilled				**Time Gap**	**Our Time**		

Fulfilled Prophecy of Jesus Recorded in the Gospel of Luke

FOR THE DESTRUCTION OF HEROD'S TEMPLE, DANIEL'S "TIME-oriented" prophecy hid the 14,000 days that start on Palm Sunday, then end with Roman armies burning down the temple.

To help explain the relationship of these two 14,000 days sequences, you must read further into what Luke wrote. We discover that Jesus' words link this final 14,000 days sequence to the second coming of Christ as the following verses support.

> "... upon the earth dismay among nations, in perplexity at the roaring of the sea and the waves, men fainting from fear and the expectation of things which are coming upon the world; for the powers of the heavens will be shaken. And then will they see THE SIGN OF THE SON OF MAN COMING IN A CLOUD with power and great glory" (Luke 21:25-27).

At this point, please remember that Luke did not frame the question (in Luke 21:7) about the second coming of Christ. For sure, Luke wrote the question to ask about the destruction of the temple. However, Jesus made the statement that links two periods with precisely 14,000 days aligned to the foretold events.

So that you understand, the destruction of Herod's Temple was foretold in the time-based prophecy in Daniel.

In the modern era, the final 14,000th day is the Feast of Trumpets. How can the Feast of Trumpets be prophetic?

We learn from Christian history that the feasts recorded in the Old Testament are prophetic of New Testament events. For example, the Passover is prophetic of Jesus dying on the cross as a sacrificial offering. Our 20/20 hindsight links Daniel's time-based texts for the time of Messiah appearing with the passage related to atonement, which means a sacrificial offering as follows:

"^{24}...to finish the transgression, to make an end of sin, to make atonement for iniquity, to bring in everlasting righteousness, to seal up vision and prophecy, and to anoint the most holy place" (Dan. 9:24).

The information was in the prophetic texts, but people did not understand.

After the Passover, the second feast always happens on Sunday, which is the Feast of First Fruits. Jesus rose from the dead on the Feast of First Fruits.

The third feast that happens fifty days later is the Feast of Weeks. We call this the day of Pentecost, which is the day the Holy Spirit anointed the believers at Jerusalem, as described in the book of Acts. So, these feasts are prophetic of New Testament events. At the first coming of the Messiah, New Testament events happened on the Springtime feast days.

At the second coming of the Messiah, the autumn feasts could very well reveal the exact days of massive change. The very first autumn feast is the Feast of Trumpets. Since the second coming is associated with the blowing of trumpets,[4] why is the 14,000th day on October 4, 2005, aligned to the prophecy in Luke 21:24b?

Some may ask, "Why didn't Jesus return on the Feast of Trumpets on October 4, 2005?"

The answer to this question comes from the gospels of Matthew and Mark. We will find a similar pattern to what Luke wrote. By this, I mean, a question that requires a single sign. However, the sign recorded in Matthew and Mark is very different from the gospel of Luke sign.

End Times Sequence in the Synoptic Gospels

Matthew and Mark frame similar questions. Consider what Matthew wrote.

> "... when will these things be, and what will be the sign of Your coming, and of the end of the age" (Matt. 24:3)

Matthew's question inquires for a single sign of the second coming of Christ. In contrast, Luke did not frame a question like this. Luke did reveal that Jesus spoke about the second coming of Christ that is directly related to the single sign of Jewish people regaining control of Jerusalem.

What is the single sign in the gospels of Matthew and Mark?

In Mark and Matthew, the single sign referred to is called the abomination of desolation spoken of by the prophet Daniel. Consider the text:

> "...when you see the ABOMINATION OF DESOLATION which was spoken of through Daniel the prophet, standing in the holy place..." (Matt. 24:15).

Daniel foretold of this sign in the time-based texts. This event did not happen at the time that Romans burned down Herod's Temple on the final 14,000th day. Some people teach that these events happened in ancient history. However, this seems to be completely wrong based on the sequential nature of the time-based texts. Because the time-based documents have the "abomination" sign occurring after Roman armies destroy the temple and Jerusalem. Regardless, the abomination sign requires a temple. Also, the following event occurs after the abomination sign.

---

^{29}"...immediately after the tribulation of those days . . . the POWERS OF THE HEAVENS WILL BE SHAKEN, ^{30}and then the sign of the Son of Man COMING ON THE CLOUDS OF THE SKY with power and great glory" (Matt. 24:29-30).

The abomination event and the return of Christ did not happen in ancient history.

In my opinion, this major event will require a lot of change because a temple in Jerusalem is necessary to fulfill these prophetic texts.

People who want to rebuild the temple need to discover that Jesus' atoning death and subsequent resurrection replaced Herod's Temple by way of its destruction on the 14,000th day.

What should we expect to happen?

In our time, if Jewish people rebuild the temple, those people will have rejected the 14,000-day alignment in Daniel's prophecy. Those people will meet face to face with a highly spiritual person we often refer to as the antichrist. Currently, we live before that event.

The bottom line is that we are living in the days that Jesus answered Luke's question. The texts in Luke reveal that the Second Coming of Christ is about to take place. In other words, the Israeli conquest of Jerusalem in 1967 shows we are getting close, but not at the end.

The single sign recorded in Matthew and Mark reveals that the second coming is going to happen. Matthew wrote,

> "...the sign of Your coming, and of the end of the age" (Matt. 24:3).

Matthew's question links to the abomination of desolation in Daniel's time-based texts. These texts support a final seven-year period with the required temple constructed in Jerusalem.

What Will Happen Next?

It appears that the gospel of Luke sign gives information that we need to prepare for the Second Coming of Christ.

However, the sign described in Mark and Matthew awaits a future fulfillment.

It appears there is some time left for the church to reach out to non-Christians. People on planet earth can prepare for the unveiling of Christ when he returns with power and great glory. I have graphed the layout between the gospels of Luke, Mark,

Eternal Life - why you should expect to live forever

and Matthew so you can get the relationship for the timing of the Second Coming of Christ.

In this graph, we live in the black area with white letters of this prophetic landscape. To the left are the fulfilled texts of Jesus recorded in Luke 21:24b. The time for the blackened area is unknown. The portion on the right represents the fulfillment of the gospels of Mark and Matthew. These gospels point us back to the final seven years of Daniel's "time-oriented" prophecy.

Jewish People Regain Control of Jerusalem June 6/7, 1967 Six-Day War Jesus' Prophecy fulfilled Luke 21:7, 24b Sign that Second Coming is "about to take place"	14,000 Days	Feast of Trumpets, Tuesday, October 4, 2005 Symbolic of Second Coming of Christ	Unknown Time until Prophetic Words are Fulfilled Future Event that requires Jewish Temple on the Temple Mount in Jerusalem.	Abomination of Desolation Future Event (Date Unknown) Specific Sign of the End of the Age See Matthew 24:3, 15
Modern History Fulfilled			Time Gap	Requires Matt. 24:3, 15

Synoptic Gospels - The Sequence of Prophetic Events

I find it intriguing how well the gospels portray the return of Christ. Events have occurred that support Jesus' words came to pass as recorded in the gospel of Luke. Then there will be a time for global evangelism. However, someday, the door will be shut, and the end will come quickly.

The gospel of Luke shows that we are getting close. When the sign in the gospels of Mark and Matthew occurs, then we will know that we are approaching the very doorway of Christ's appearing again on planet earth.

Perhaps you need to be reminded that many Christians believe that there will be a rapture of the believers before a time of great trouble begins. This event would likely occur in the blacked region of the graph on an unknown day. I recently watched a fictional movie about this event, Left Behind!

Spiritual deceivers will abound during the end-time scenarios explained in the Bible. This means you need to learn more about the spiritual realm. You may want to consider book four in the Expect to Live Forever book series, Secrets – never heard until now – of the Spiritual Realm.

20

DO THE TIME-BASED TEXTS FORETELL FUTURE EVENTS?

WHAT IS THE VALUE OF THIS RESEARCH

IF YOU TAKE THE TIME TO INVESTIGATE, YOU WILL FIND THAT THE precise dates for the time-based texts are accurate. You also must use the guidelines from the biblical passages. If you reject the biblical passages, then you will not agree with the conclusion that these texts are from a transcendent Being. Hereafter is a quick overview to aid in your investigation.

Biblical Numbers with Repeatability

You must use the number seven called out in the time-based texts. The idea of seven years being a sabbatical is required. We also have seven days each week. In math, when you multiply numbers, the use of the number seven means you will always get the same day of the week.

The use of 360 days in a year is required. The Bible refers to the 360 value in at least four passages in the Bible. We have found that when you use the 360 value, we unveil exactly 14,000 days hidden in the passages around the lives of Moses and Joshua. The repeatability occurs the second time in Daniel's time-based texts as we find 14,000 days from Palm Sunday to the day that Roman soldiers burned down Herod's Temple.

When were the Texts Recorded on Paper?

The differing views range from 530 BC to 165 BC. If you accept data analysis that compares the Dead Sea Scrolls to the text in Daniel, the 530 BC date is more credible. If you use the opinion of a rationalist, then the 165 BC date can be used. Regardless, the destruction of Herod's Temple takes place from 235 to 600 years into the future. Therefore, the texts foretell the destruction of Herod's Temple.

When we uncover precisely 14,000 days that terminate on the day that Romans burn down Herod's Temple, the texts take on new meaning. In the Bible, the Jewish people used the temples only for sacrificial services. Daniel's time-based writings link the Messiah's death and implied resurrection with the burning down of Herod's Temple. We logically conclude that God orchestrated temple destruction events.

Scientific Date for Artaxerxes' Decree to Rebuild Jerusalem

The decree of Sunday, March 16, 445 BC, is credible for the following reasons.

- Sunday is required based on the biblical texts that use the number seven. The events must occur on the same day of the week.

- Elephantine papyri point to the Springtime of 445 BC. Scholars that claim this occurred in 444 BC blame the scribes for human error. Also, they cannot scientifically date the Cairo Sandstone Stele.
- Babylonian astronomical texts from the third century BC verify the Mishnah's statements about calendar variation. This data validates the precise dates for the Elephantine papyri as well as the Cairo Sandstone Stele. In contrast, without this data, scholars blame the scribes for human error based on weak presuppositions.

Counter Arguments Summary:

1. Scholars choose to reject the Mishnah based on the modern era's understanding of astronomy. This group would believe that the Talmud outranks the Mishnah. The problem is that astronomical data from the third century BC validates the Mishnah. The Talmud contains the views of people who used the calculated Hebrew calendar that did away with the actual practice of the calendar when Jesus walked the earth. Modern era scholars have created a false paradigm related to the actual practice of the Hebrew calendar.
2. Sir Robert Anderson claimed this occurred on Friday, March 14, 445 BC. However, we find this to be an error since the time from this date to Palm Sunday would be 173,882 days. Also, the time-based texts require the same day of the week, which must be a Sunday.

JOHN ZACHARY

Scientific Date for Destroying Herod's Temple

Roman soldiers burned down Herod's Temple on Sunday, August 5, AD 70, which is credible for the following reasons.

- Sunday is the day of the week used by eyewitness reports.
- Ten days since a new moon as it burned down on Ab 10.
- Evidence and logic point to only one answer, which is Sunday, August 5, AD 70.
- Dr. Jack Finegan calculated the same time as me based on evidence, astronomy, and logic.

Biblical Date for Jesus Entering Jerusalem as the Messiah

Palm Sunday comes from the time-based text calculations, which means you must accept the biblical math in the texts. For sure, the use of science cannot isolate the date for Palm Sunday. However, this calculated date separates the hidden 14,000 days from Palm Sunday to the destruction of Herod's Temple.

- Sunday is the day of the week based on the time-based texts use of the number seven.
- Repeatability of the 360 days per year unveils exactly 14,000 days hid inside the time-based passages.
- The biblical calculation requires adding 173,880 days to the decree to rebuild Jerusalem on Sunday, March 16, 445 BC.
- Sir Robert Anderson calculated the 173,880 days as published in <u>The Coming Prince</u> released in 1894.

Counter Arguments Summary:

- Scholars may not like using the 360 days per year value. In truth, this means the scholars choose to ignore biblical math. If that is you, then you will never find the 14,000 days hidden below the surface of biblical texts. In truth, the value of 360 gives repeatable results.
- Jesus had to die in a year with the full moon on a Thursday or Friday. If you choose this viewpoint, then you have ignored the texts in the Mishnah that are corroborated by Babylonian astronomical data from the third century BC. I prefer the data over this viewpoint. The full moon on Monday is credible for the year of Jesus' crucifixion.

What is the Value of this Research

Consider the unique information about the spiritual realm influencing human events. In chapter two, I wrote the premise and critical questions. *"Will we observe spiritual beings influencing human events in our universe?"*

The logical answer is that we will observe such things. The result of scientific dating the time-based texts supports that we have found the ultimate effect from a transcendent Being. Why would this be okay?

To begin, Jesus claims to be the Son of God[1], the second member of the Trinity. The time-based texts foretell the precise date of events related to this Messiah, who is God-Almighty. Daniel prophesied of Jesus' entry into human activities, so we would know of the Messiah and the influence of events to reveal spiritual truth to humanity.

The 14,000 days from Palm Sunday to the burning down of Herod's Temple are at the pinnacle of God-like influence. Why?

Please understand that the time-based texts do not give us math to make this calculation ahead of the two events. We see repeatability with the burning down of both Solomon's Temple and Herod's Temple on the final 14,000th day. Logic leads to the conclusion that God orchestrated temple destruction events. Israel built those edifices for sacrificial services.

Consider that Jesus as Messiah is the ultimate sacrifice, and the purpose of the time-based texts is that the Messiah will die as a sacrificial offering. We perceive that the transcendent Being, God, controls temple destruction events. These foretold events come from a Being that knows all future incidents to an exact date. It appears that incidents with high-level spiritual meaning require God-like control of some humans to make the events occur on an exact day as foretold from centuries earlier. Burning down Herod's Temple on the final 14,000th day supports Jesus' sacrificial death and his supernatural resurrection. According to New Testament theology, eternal life requires placing your trust in Jesus' death on the cross and his supernatural rising from the dead. The time-based prophecy rubber-stamps Jesus as Messiah with the controlled destruction of Herod's Temple on the final 14,000th day. We perceive the reality of eternal life.

How to use this information

The Qur'an has no passages that foretell of specific future events that will occur on an exact date, centuries into the future. We unveil a primary problem with the Qur'an.

Just as important, the Christian faith and the Islamic faith have diametrically opposite views of Jesus. The central message of the Bible is that Messiah redeems people through a sacrificial offering and supernatural resurrection. Destruction of Herod's Temple on the final 14,000th day magnifies the work of Christ, who did away with earthly temples. Profoundly, Daniel's time-based text links the Messiah's sacrifice to Herod's Temple used for sacrifice.

Jesus boldly proclaimed that the only way into eternal life was to place your faith (trust) in himself. Consider that New Testament texts teach that Jesus atoned for you by dying on the cross, then rising from the dead.

In contrast, the Qur'an teaches that Jesus never got on the cross. The message of salvation in the Qur'an is that a Muslim must become righteous by good works alone. What do these diametrical opposite views reveal about the source of the Qur'an?

To begin, we perceive that the spiritual source of the Qur'an cannot foretell future events with great detail, centuries into the future. We logically conclude that the source, Allah, is not a transcendent Being!

Then the directly opposite message of salvation revealed in the Qur'an supports that the source is a religious deceiver[2] whose purpose is to draw you away from Jesus.

The evangelism of Muslims should become a high-level activity of the church. However, will the church respond?

Pantheistic religions do not have the equal of this valuable information.

Pantheistic religions have no transcendent Beings. The information herein supports that Pantheism is false.

The evangelism of Pantheist religions should become a high-level activity of the church. However, will the church respond?

Please consider directing people to the following website, which will be changed to help people of all faiths begin to gain an understanding of this valuable information.

www.expect-to-live-forever.com

www.expecttoliveforever.com

21

GLOBAL-WIDE CHRISTIAN REVIVAL

Do you believe there will be a global-wide Christian revival?

Then please go to the Expect to Live Forever website shown below. You will discover a method for mass education and evangelism that stops Islamic growth. When a community learns that the Qur'an comes from a source locked inside space-time, no logical person will become a Muslim.

When people stop becoming Muslims, then followers of Islam want to know why. Some Muslims will decide to leave Islam for the reality of Christ. You can read the evidence in support of this viewpoint in chapter twenty, subheading: **How to use this information.**

If you want to get involved, you must get onto the email system on the website below. Also, you will only learn this method by enrolling.

www.expect-to-live-forever.com

PLEASE REVIEW MY BOOK

BY REVIEWING MY BOOK, YOU CAN MAKE A DIFFERENCE!

Thank you for reading the second book in the Expect to Live Forever book series.

Reviews are the most powerful tool when it comes to getting attention for the Expect to Live Forever books.

I would love to hear what you have to say. Other people do gain value from your insights. Please leave an honest review on Amazon, letting me know what you thought of this book. Imagine, your review could help others receive eternal life!

Thanks so much! I pray that you enter the world of miracles and dreams come true, forever!

John Zachary

ACKNOWLEDGMENT

I am grateful to Sir Robert Anderson (1841 – 1918). He authored The Coming Prince, published in 1894.

I am thankful for Pastor Frank, who cheered and guided me as I wrote these materials.

EXPECT TO LIVE FOREVER BOOK SERIES

1. Secrets – *never heard until now* – of the Book of Revelation
2. Eternal Life - *Why you should expect to live forever!*
3. Beyond - *The Coming Prince by Sir Robert Anderson*
4. Secrets – *never heard until now* – of the Spiritual Realm
5. More books with cutting edge materials are coming! To keep up to date, consider getting on the email list by going to the following website contact:

www.expect-to-live-forever.com

www.expecttoliveforever.com

I look forward to hearing from you. You can get information for new releases as well as ministry opportunities at the Expect to Live Forever website.

ABOUT THE AUTHOR

I wrote this book based on my keen interest in both science and religion. To a large extent, how they support one another. For instance, the Bible has texts that foretell of specific events that will occur on an exact day, hundreds of years into the future. No human being could do this!

Evidence in support of this viewpoint comes from the Dead Sea Scrolls. Those documents verify the texts existed long before the future events. Then I use science to find the precise date of each foretold event. Words that predict the future like this must come from a Source that transcends time. This method requires both the texts in the Bible and science to find the precise dates of the activities.

I have studied these methods for almost four decades. My degrees include History-Education and Mathematics.

I worked as a Reliability Engineer for about twenty years in aerospace. My job was to make the products highly reliable. When I solved a problem that three other engineers could not fix, my career outlook enlarged. I worked my way to the top of the engineering department.

I enjoyed improving hardware. A good example was a design change that combined nineteen parts into a single piece. This design change saves the US government about twenty-million dollars over ten years.

Another design change improved rocket launches. Instead of two failures in one-hundred launches, this design change made it impossible to fail. The basis for these advances was my ability to work with data. I found the data. Then I studied it to create tests that improved the hardware.

In this book, I apply these skills to analyze biblical prophecy. However, there is more to this than using my engineering skills. Perhaps you would ask, *"Why would an engineer be caught up in studying biblical prophecy?"*

If you want to know why I am captivated with this subject, please read book one of the Expect to Live Forever book series, Secrets - never heard until now - of the Book of Revelation. You will find Twilight Zone type stories that support, "You should expect to live forever!"

BIBLIOGRAPHY

1. Anderson, Sir Robert, The Coming Prince, published 1894 (Christian Classic, now corroborated by ancient astronomy without scholar's paradigm based presuppositions. Scholars often blame ancient scribes for human error to support their weak viewpoints.)
2. Archer, Gleason L., Encyclopedia of Bible Difficulties, p. 282 - 283, published 1982, source of the 530 BC composition date based on data.
3. Davies, Paul (12 April 2003). "A Brief History of the Multiverse." The New York Times.
4. www.closertotruth.com/ "Paul Davies – Are There Multiverses?"
5. Denis Dutton, Sky and Telescope, Sept. 1984, Paul Davies on the Existence of God, pp 229-30 denisdutton.com/davies_review.htm
6. Ferguson, Kitty (2011). Stephen Hawking: His Life and Work. p. 96. Transworld. ISBN 978-1-4481-1047-6.

Bibliography

7. Finegan, Jack, <u>Handbook of Biblical Chronology</u> Table 51, p. 108, (c) 1998.
8. Hazen Robert M., George Mason University and Carnegie Institution of Washington, 24 lectures on "<u>Origins of Life</u>" research. Distributed by The Great Courses in Chantilly, VA 20151-2299 www.thegreatcourses.com (800) 832-2412
9. Heeren, Fred, "<u>Show Me God: what the message from space is telling us about God</u>," DayStar Publications, copyright 1995, ISBN 1-885849-52-4.
10. Hemer, Colin J., <u>The Book of Act in the Setting of Hellenistic History</u>, Published 1990 by Eisenbrauns
11. Horn, Seigfried, Wood, Lynn, <u>The Chronology of Ezra 7</u>, 2nd Edition Revised
12. https://www.iceagenow.info/global-temperature-fluctuated-long-industrial-revolution-graphic/
13. http://numerical.recipes/julian.html - A Julian Day and Civil Date Calculator.
14. https://sacred-texts.com/jud/josephus/war-6.htm Scroll down to chapter four, paragraph 5.
15. https://theosophical.wordpress.com/2009/12/19/signature-in-the-cell-part-4-assessing-the-chance-hypothesis-for-the-origin-of-life/
16. https://en.wikipedia.org/wiki/English_words_first_attested_in_Chaucer.
17. https://www.biblestudytools.com/dictionary/atonement/
18. https://en.wikipedia.org/wiki/Hermann_von_Helmholtz
19. Josephus, <u>Antiquities of the Jews</u>
20. Meyer, Stephen C., "<u>Signature in the Cell: DNA and the</u>

Bibliography

<ul style="list-style: none;"><li>Evidence for Intelligent Design," Harper Collins, copyright 2009 by Stephen Meyer.</li></ul>

21. Neusner, Jacob, The Mishnah, a new translation, Yale University Press 1988.
22. NOVA Documentary "Runaway Universe" pbs.org
23. Parker, R.A., Dubberstein, W.H., Babylonian Chronology 626 BC to AD 75, Brown University Press, 1956.
24. Pauken, Michael, Thermodynamics for Dummies, Part II, p. 4, "Employing the Laws of Thermodynamics" quoted statement.
25. Principe, Lawrence M., John Hopkins University, 12 lectures on "Science and Religion." Distributed by The Great Courses in Chantilly, VA 20151-2299 www.thegreatcourses.com (800) 832-2412.
26. Stenger, Victor J., God – the failed hypothesis
27. Sachs, Abraham J., Steele, John M., Hunger, Hermann, Astronomical Diaries and Related Texts from Babylonia, Vol. V, Lunar and Planetary Texts, 2001.
28. Stern, Sacha, The Babylonian Calendar at Elephantine, Table I, row C13. based on Habelt, Dr. Rudolph, Zeitschrift fur Papyrologie und Epigraphic, Bd 130 (2000)
29. Sukenik, E.L., "The Earliest Records of Christianity," The American Journal of Archaeology, Vol. LI, No. 4, 1947, p27-29.

NOTES

1. Can we Know that God Exists?

1. A subatomic particle
2. Atheism is a *belief system* that denies the existence of God. In truth, Atheism is a belief system about ultimate reality. For sure, science supports that Atheism is false.
3. Website or search for the name of this scientist: https://en.wikipedia.org/wiki/Hermann_von_Helmholtz
4. Pauken, Michael, Thermodynamics for Dummies. Biography of Dr. Michael Pauken: Senior mechanical engineer at NASA, who develops spacecraft thermal control systems and balloon systems for exploring other planets in the solar system.

2. The Ultimate Premise and Question

1. Entropy means disorder or that the universe will come to an end.
2. Atheist scientists continue to claim that a natural world may perhaps exist that has always existed. However, known scientific data does not support that viewpoint. The Atheist worldview requires blind faith in unproven theoretical views. Atheist believers want people to believe in a natural universe that is God-like. To accept this requires enormous levels of blind faith. In contrast, the Expect to Live Forever book series provides evidence and logic to support supernatural influence over human events. If you want to investigate at a deeper level, please read Beyond - The Coming Prince by Sir Robert Anderson.
3. Entropy means disorder or that the universe will come to an end.

3. What existed before the Universe Began?

1. Some people will claim that the Supreme Being is not personal.

Notes

4. Atheism: Its Quest for an Eternal Natural Universe

1. Paradigm is a widely accepted set of beliefs or concepts. Paradigms can be utterly false until evidence reveals the viewpoint to be erroneous. As a result, changing a paradigm can take many years, even decades or millennia, to reverse. The use of the word, 'paradigm' is an excellent way of revealing the probability that a person may exist in a brainwashed state.
2. Consider that Albert Einstein used his mind to find advanced ideas such as special relativity and general relativity. However, Einstein changed his mind based on data from the real world. Now think about Atheist thinkers that want you to believe that they can imagine that God is not required. Will data change their minds too?
3. NOVA Documentary "Runaway Universe" pbs.org
4. The phrase "scientific atheism" is an oxymoron. People in this belief system like to refer to themselves this way. However, the correct terminology is "philosophical atheism," as described by Professor Lawrence M. Principe, John Hopkins University, 12 lectures on "Science and Religion." Distributed by The Great Courses in Chantilly, VA 20151-2299 www.thegreatcourses.com (800) 832-2412. In general, Atheism has a presupposition that there is no supernatural domain or spiritual realm. To maintain this viewpoint, please understand how atheists think. "Atheists simply do not care if they are wrong!"
5. Consider Stephen Hawking's circular reasoning to arrive at the belief that God does not exist. Mr. Hawking reasoned the existence of a so-called "no-boundary universe" that is infinite and eternal. In Atheist thought, this would be the natural universe that is equal to God. Then this eternal and infinite universe must create multitudes of other worlds to explain the only observable universe in which we exist. Are there any flaws in Mr. Hawking's thoughts?

    If the no-boundary universe exists, then each time it creates another world, some of its energy/mass will go away. In contrast with energy/mass, time into eternity past will not be affected by the creation of other worlds. Time becomes the nemesis of the no-boundary universe hypothesis. We perceive that the eternity past is credible, but infinite energy/mass is not probable. The flaw in the no-boundary assumption is the endless time going backward forever. Losses of energy/mass across eternity past will not hold up. The fact of losing energy/mass validates the imagined idea of infinite energy/mass is false. Put a different way; no one will debate time into eternity past. However, the loss of energy/mass into eternity past is debatable.

Notes

As a result, the no-boundary universe will be dying. Based on logic, the no-boundary world does have limits. Only time is limitless. We perceive the ever-present flaw of human error based on imagination! Also, we have absolutely zero data to support what Stephen Hawking proposed as a replacement for God.

6. Ferguson, Kitty (2011). Stephen Hawking: His Life and Work. p 96. Transworld. ISBN 978-1-4481-1047-6.
7. Entropy means disorder or that the universe will come to an end.
8. Berlinski, David, The Devil's Delusion: Atheism and its Scientific Pretensions, pp 110-111. Refers to physicist Paul Davies and cosmologist Fred Hoyle.
9. Davies, Paul (12 April 2003). "A Brief History of the Multiverse" The New York Times.
10. Wikipedia, an article on "Hubble's law" with mathematics used to explain the distance to the "Hubble Length," which is the distance where stars cross the threshold of moving away from earth at faster than light speed. Perhaps you may find it interesting that the stars position themselves inside space. However, it is space that is increasing its expansion rate beyond light speed. Space, not the stars, is the vehicle of acceleration.
11. Clarification: The calculated distance across the universe is between ninety-billion and 100 billion light-years. This distance is from the time of the initial Big Bang Event. If you took a photo of all the matter in the universe today, all materials that are beyond fifteen billion light-years from the earth are beyond sight. However, we can observe light from stars before they cross the Hubble length. Therefore, we can see stars that existed less than fifteen billion light-years ago. When a star exceeds fifteen billion light-years distance from the earth, then that star will never be observed from that time forward.
12. Entropy and the Second Law of Thermodynamics in Particle and Nuclear Interactions, San Jose State University. In contrast, the Atheist worldview propounds that the source of our universe came from pre-existing matter consisting of a soup of quantum particles. This viewpoint and its presuppositions appear to be false based on known scientific data. It is a fact that all subatomic particles cannot sustain themselves or bypass the principle of increasing entropy.
13. Consider a thought-provoking viewpoint on "natural gods" based on the case for Pantheism. In the Pantheist religions, the existing universe creates the gods. Therefore, this is a proper application of Dawkins's thought since he wants to know, "Who created god?" In reality, this is a foundational reason that reveals Pantheistic beliefs are flawed.
14. If you are intellectually inclined, you will investigate both resources referenced below since they are on opposite sides of the fence. Otherwise, if

you are not inclined to finding spiritual truth, you will ignore this information (my opinion)! People who think evolution explains everything base their viewpoint on data of simple cells on the earth over 3.7 billion years ago. Those people choose to scoff at the research done by Stephen Meyer. I list two sources of information within this footnote.

Stephen Meyer's research and materials show a good example that explains the complexity of these problems. Discover why it will be difficult for scientists to find an answer for amino acids self-assembling into living cells. Meyer, Stephen C., "Signature in the Cell: DNA and the Evidence for Intelligent Design", Harper Collins, copyright 2009. I recommend getting the DVD as it will explain the issues at hand and the complexity of living cells that are mind-boggling. Also, the probabilities are accurate due to the necessity of each amino acid sequence required to assemble a protein molecule.

http://theosophical.wordpress.com/2009/12/19/signature-in-the-cell-part-4-assessing-the-chance-hypothesis-for-the-origin-of-life/

Look into an excellent secular example of such research in "Origins of Life," written by Professor Hazen. Hazen scientifically documents how natural processes will create amino acids. However, this claim is feeble because cellular life requires the amino acids to combine into complex proteins and super complex molecules such as RNA or DNA. The "key problem" is that amino acids must self-organize to form a viable, simple cell. Professor Hazen even admits that scientific investigation may never discover how life could have originated by chance events. People who claim life originated by chance do not use experimental data.

Professor Robert M. Hazen, George Mason University and Carnegie Institution of Washington, 24 lectures on "Origins of Life" research. Distributed by The Great Courses in Chantilly, VA 20151-2299 www.thegreatcourses.com (800) 832-2412

15. Based on scientific dating Daniel's "time-oriented" prophecy. When the end times come, spiritual awareness will increase to the highest heights! "… conceal these words and seal up the book until the end of time; many will go back and forth, and knowledge will increase" (Dan. 12:4).

5. Pantheism:

1. Or the contrasting view that the Universe is entirely spiritual, meaning our perception of the physical universe is not valid. This idea is purely philosophical.

Notes

2. Denis Dutton, Sky and Telescope, Sept. 1984, Paul Davies on the Existence of God, pp 229-30 denisdutton.com/davies_review.htm
3. Perhaps I need to remind you that in chapter one, I wrote about three methods that unveil the spiritual world, just like the air we breathe! The first method comes from using the exact science of astronomy to date Daniel's time-based prophetic texts scientifically.
4. WMAP satellite data calculations by NASA scientists now claim that the universe began 13.78 billion years ago, with the error being about plus or minus 50 million years. Therefore, the maximum time value based on scientific data would be 13.83 billion years ago for the creation of the observable natural universe.
5. Heeren, Fred, "Show Me God: what the message from space is telling us about God," DayStar Publications, copyright 1995, ISBN 1-885849-52-4. Documents deuterium on page 161 and the fact of high temperatures occurred in the primordial high temperatures of the Big Bang event. To form deuterium requires a temperature higher than one-billion degrees Celsius (same as 1.8 billion degrees Fahrenheit).
6. Entropy means disorder or that the universe will come to an end.

6. Monotheism:

1. If you began reading in chapter six, I describe and define the Magic-Verse in the fourth chapter.
2. "...flesh and blood cannot inherit the kingdom of God, nor does the perishable inherit the imperishable" (1 Cor. 15:50b).
3. I find it interesting that Atheists think the biblical texts should foretell of events observed in the modern era exactly as the Atheist alone would accept as proof. For example, Atheist Victor Stenger propounds that the Bible would predict of men on the moon playing golf [ref: Stenger, Victor J., God – the failed hypothesis, p 176]. I find it just as intriguing that Professor Stenger fails to acknowledge the environment of the book of Revelation recorded in Rev. 11:9-10, which requires satellite technology. Or consider the destruction of Damascus in a single night from a recorded source around twenty-seven centuries ago [ref: Secrets – never heard until now – of the Book of Revelation, p 68-70]. The main point of this footnote is that God does not foretell events simply to prove that God knows the future. Instead, the "time-oriented" texts only predict future events that reveal spiritual truth. Please consider the fact that we have no data to support the atheist hypothesis that God does not exist. That data will never exist based on the multiverse hypothesis. Professor Stenger should retitle his book to Atheism – the failed belief!

Notes

4. Consider reading book one, Secrets - never heard until now - of the Book of Revelation as it gives insight into why I write the Expect to Live Forever book series.

8. When Was Daniel Written?

1. https://en.wikipedia.org/wiki/English_words_first_attested_in_Chaucer. Select words found in Chaucer's manuscripts that we use today, which reveals changes in the English language since the 14th century.
2. "Now that we have at least one fairly extensive Midrash originally composed in the third century B.C. Aramaic and several sectarian documents in second century Hebrew, it has become possible to perform a careful linguistic comparison of the Aramaic and Hebrew chapters of Daniel and these unquestionably third or second century BC documents, which were close to the era of the Maccabean struggle.

    If Daniel had in fact been composed in the 160s, these Qumran manuscripts should have exhibited just about the same general characteristics as Daniel in the matter of vocabulary, morphology, and syntax. Yet the actual test results show that Daniel 2 – 7 is linguistically older than the Genesis Apocryphon by several centuries. Hence these chapters could not have been composed as late as the second century or the third century, but rather - based on purely philological grounds – they have to be dated in the fifth or late sixth century; and they must have been composed in the eastern sector of the Aramaic speaking world (such as Babylon), rather than in Palestine (as the late date theory requires).

    Archer, Gleason L., Encyclopedia of Bible Difficulties, published 1982, p 283. is the source of the 530 BC composition date based on test data of documents with known composition dates.
3. Josephus, Antiquities of the Jews, book eleven, chapter eight.
4. Called the Florilegium.

10. Upon what day of the week did each event occur?

1. Scientific dates require positions of the sun, the moon, and the earth. The appearance of the Messiah comes from the biblical math detailed in the time-based prophecy.

Notes

11. Repeatability of the Biblical Constants

1. Based on Sir Robert Anderson, The Coming Prince, which aligns positions of the sun, the moon, and the earth across a period of 476 years. Mr. Anderson gave us the first viable answer. This chapter shows another example.
2. The story of Noah in the Ark: Gen. 7:11, 24, Gen. 8:3-4 In these verses, 150 days are equal to 5 months. 150 days/5 months = 30 days/month.
3. Numbers 20:29, Deut. 21:13, 34:8.
4. The calculation includes the beginning date of Nisan 1. However, the Israelites did not leave Egypt until Nisan 15. The entire calculation includes using Nisan 1 as the beginning date, which results in a period of 14,000 days from leaving Mount Sinai to crossing the Jordan River, which is the exact day they entered into the Promised Land.

12. Insights from the Mystical 14,000 Days

1. An easy way to do this is with the website: http://numerical.recipes/julian.html. Put your birthday into the top cells. In the second row, type in the number 14,000. You will find your 14,000th day on the bottom row. The software gives you the day of the week. Also, make sure you use AD, not BC.
2. I need to clarify that the thirty-eight years are actually added to the first complete year. We would actually be in the thirty-ninth year, month two, the twentieth day. I suspect this would confuse some people. This explains why the 14,000 days are mystical as you read the paragraphs after this footnote.
3. The following verse supports that a 14,000-day time span may have occurred or at worst case, it is very close to the 14,000 days.

    "...the time that it took for us to come from Kadesh-barnea until we crossed over the brook Zered, was thirty-eight years; until all the generation of the men of war perished from within the camp, as the Lord had sworn to them" (Deut. 2:14).

    We see in the verse above that thirty-eight years passed. Therefore, it is getting close to the 14,000-day value. However, we cannot find the exact date on which the Jewish people were forbidden to enter the Promised Land because of their unbelief. Also, we cannot find the precise time that they crossed the Brook Zered referenced in that verse. Therefore, we can conclude that there is no possibility of finding precisely 14,000 days at the time of Moses and Joshua for the generation that died in the desert.

Notes

However, you may find it intriguing that Jewish oral tradition declares that the judgment on the generation that died in the desert began on Ab 9. By adding four months to this date would equate to crossing the Brook Zered about Kislev 9. The estimation comes near 14,000 days. We find that Aaron, the High Priest, died on Ab 1 in the fortieth year as recorded in Num. 33:39. After Aaron's death, the nation mourned for thirty days, as recorded in Num. 20:29. Hence, they would have crossed the Brook Zered after Elul 1. We find that Num. 21:1-12 includes several events that would likely require a few months to occur after Elul 1. Therefore, we can only estimate that close to 14,000 days transpired from the judgment at Kadesh Barnea to the final death observed at crossing the Brook Zered. Also, Moses' death does not apply since he was not under the death verdict given at Kadesh Barnea.

13. Jewish Temples: Purpose and Controlled Destruction

1. Two eyewitness accounts: Firstly, Jer. 52:12-13 - "... on the tenth day of the fifth month [Ab]... he burned down the house of the Lord..."

    Second, Josephus' eyewitness account of Romans burning down Herod's Temple records, "...it was the tenth day of the month Lous [Ab], upon which it was formerly burnt by the king of Babylon." [Josephus, Wars of the Jews, 6.4.5].
2. Finegan, Jack, Handbook of Biblical Chronology, revised edition, p. 107, paragraph 203.
3. Ibid.
4. Ibid.
5. Based on John 14:6, Jesus boldly claimed that he alone could give you eternal life. This viewpoint finds support in Romans 10:9-11. These scriptures reveal that Jesus' sacrificial death and supernatural resurrection give you eternal life. Destruction of Herod's Temple on the final 14,000th day supports that the earthly sacrificial system ended because Jesus' death and supernatural resurrection replaced it. We find corroboration of this viewpoint because the destruction of Herod's Temple was foretold as an event many centuries ahead of time in the time-based texts. Also, the central theme of the time-based passages recorded in Daniel 9:24 reveals that the Messiah would atone for your sins.
6. https://www.biblestudytools.com/dictionary/atonement/

    The referenced website has much more about the biblical use of the word atonement should you want to dig deeper.
7. Refer to Exodus 12:3-6.

8. Refer to Jeremiah 52:12-13, the day foreign armies burned down Solomon's Temple is repeated by foreign troops burning down Herod's Temple.
9. Consider that verse twenty-four reveals six results from the predicted events, which explain why burning down both Solomon's Temple and Herod's Temple on the 14,000th-day links up with God-like controls.

    To finish transgression (to stop wrongdoing)

    To make an end of sin (to wipe away every sin)

    To make atonement for iniquity (a sacrificial offering for sin, so you become spotless)

    To bring in everlasting righteousness (in Christian theology, righteousness given as a gift to each person who places their trust (faith) in Jesus, the Messiah).

    To seal up vision and prophecy (pointing to the pinnacle of Old Covenant prophecies fulfilled in the Messianic character). Also, that subsequent beliefs that claim to receive new revelations are false. Any so-called holy writ composed after the book of Revelation will be for spiritual deception. The ideal example is the Qur'an, which claims that Jesus did not even get on the cross.

    To anoint the most holy place (literally, the site called the holy of holies in the earthly temple where the Jewish high priest sprinkles the blood of a lamb. In contrast, the Messiah goes into the holy of holies in a temple that is not of this world, which is an everlasting temple. If you want to dig deeper, you will find eternal life in a supernatural universe. Refer to Hebrews 9:1-26). No earthly priest could accomplish this requirement.

    These biblical texts portray a spiritual temple that is not of this world. In contrast, Daniel's "time-oriented" passages foretell the destruction of Herod's Temple. Is there a link between the physical temple at Jerusalem to the supernatural temple where Messiah is the high priest in whom you can place your trust (faith) to receive eternal life freely. Consider studying Romans 10:9-11, Rev. 21:22, and Hebrews 9:1-26.

14. Burning Down Solomon's Temple

1. In the synoptic gospels of Matthew, Mark, and Luke, Jesus died on the fifteenth day. In Egypt, the firstborn also passed away on the fifteenth. Hence, the suggestion that Jesus died in the place of humanity. However, the gospel of John portrays Jesus on the cross occurred on the fourteenth day, which repeats the book of Exodus. John wrote that gospel from a theological perspective.

    The difference in dates could be actual because the sect called the

Sadducees practiced their calendar different from the sect called the Pharisees. For example, the day of Pentecost always happened on Sivan 6 for the Sadducees (refer to Ex. 19:1-11) on any day of the week. In contrast, the Pharisees followed Lev. 23:4-16, with Pentecost always occurring on a Sunday.
2. Leviticus 26 has the prophetic words about the Judgement of Israel. By the seventh century BC, the people were worshipping idols. The result of rejecting the Deliverer from Egypt leads to destruction of the Promised Land. Most of the Jewish people were transported to Babylon and lived in captivity.
3. These are the actual 14,000 days that repeat each year since foreign soldiers burned down Solomon's Temple. When you back up thirty-eight years, it is 624 BC. There is no event listed in the biblical texts for Nisan 10 that year. However, we do know that King Josiah was purifying the nation by destroying the idols. King Josiah would have selected a lamb on Nisan 10, 624 BC.
4. "...The fast of the fourth, the fast of the fifth, the fast of the seventh, and the fast of the tenth months will become joy, gladness, and cheerful feasts for the house of Judah; so love truth and peace" (Zech. 8:19).

15. Scientific Date for the Beginning of the Prophecy

1. Some biblical scholars choose to teach that Ezra 4 has this decree. However, the only order recorded in Ezra 4:21 stopped the building. The most crucial thing in Ezra 4 is that no decree with a known date was issued to rebuild Jerusalem. Another fact is that Daniel's time-based prophecy foretells that when the decree begins to rebuild Jerusalem, the construction will happen during political and military distress. Daniel's texts read, "...it will be built again, with plaza and moat, even in times of distress." Only the book of Nehemiah meets these requirements.
2. Some people may think this is biased. Although it is biased to accept the biblical texts, the positions of the sun, the moon, and the earth line up with the Sunday limitation, which unveils the 14,000 days sequences. If the words are genuinely from beyond space-time, we are required to accept the simple biblical math.
3. Select texts from the source material: "[Month V]II, ... the 3rd, equinox; ... Night of the 15th, 10° after sunset, the moon made an eclipse" NASA shows the total lunar eclipse occurred after sunset as described in the text. Sachs, Abraham J., Steele, John M., Hunger, Hermann, Astronomical Diaries and Related Texts from Babylonia, Vol. V, Lunar and Planetary

Texts, LBAT 1478, Lunar Text #57, Side: Rev., Lines 1 - 4, page 194-195. NOTE: Scholar's comments on page 194. Also, the years use the astronomy calendar that uses year 0. In this case, year 424/3 equates to year 425/4 BC in common era calendars. Equinox on Sept 28 with Tishri 3 beginning at sunset on Sept. 27. A total lunar eclipse on Oct. 9 with Tishri 15 beginning at sunset.
4. Sachs, Abraham J., Steele, John M., Hunger, Hermann, Astronomical Diaries and Related Texts from Babylonia, Vol. V, Lunar and Planetary Texts, LBAT 1419, BM 32234, Lunar Text #4, Side: Rev., Block 4, Line 4, page 21.
5. Stern, Sacha, The Babylonian Calendar at Elephantine, Table I, row C13. Based on Habelt, Dr. Rudolph, Zeitschrift fur Papyrologie und Epigraphic, Bd 130 (2000), pp 159-171.
6. In this graph, the red words are the new moon dates for the years 447 BC (21 days does not line up two days to Nov. 17), and 445 BC (14 days does not line up two days to Nov. 17). When sequences of new moons align to the Egyptian date, then we know we have the precise time of that legal document. There is a negative translation time for this document. In chapter eighteen, I explain "negative translation" times. Negative translation of two entire days is acceptable based on the details in chapter eighteen.

16. Biblical Date for the End of the Sixty-Ninth Week

1. We find the idea of a New Covenant in Dan. 9:26. In English translations, this is typically "cut off" or as used in the phrase, "Messiah will be cut off and have nothing" (Dan. 9:26b). The original Hebrew word "karath" means to covenant (i.e., make an alliance or bargain, originally by cutting flesh and passing between the pieces). Thus, Jesus' death on the cross is the literal fulfillment on a future date. Although the time-based texts do not refer to the resurrection, the texts infer resurrection. Consider the question. "If Messiah must die as a sacrificial offering to prove he is the Messiah, then how will Messiah be successful?" After all, a dead Messiah will never have influence. Therefore, rising from the dead is required for the Messianic personality, who makes a blood covenant and overcomes death and hades. Refer to Strong's Concordance word no. 3772. Remember the critical question; "Should we expect to observe supernatural events (miracles) or the influence of spiritual beings on events in this universe? The logical answer is "YES" based on the principle of increasing entropy.
2. Before doing the calculation, you may need to know that the Hebrew

Notes

word for "week" is seven years, which we call a Sabbatical. This Sabbatical comes from Leviticus chapter twenty-five, which is the counting for Sabbath years. There is also a period of seven sets of weeks that are in Daniel's "time-oriented" prophecy [*I interpret the initial forty-nine years segment as setting up for a new year of Jubilee. The land of Israel had been idle for seventy years, and the counting of Sabbath years had stopped.*]

In Daniel's "time-oriented" texts, we must add seven series of seven years to sixty-two series of seven years based on the quote of the words, "...seven weeks and sixty-two weeks..." The following equation sets up this concept.

(7 x 7 years) + (62 x 7 years)

We simplify to the following:

(7 + 62) x 7 years

Therefore,

69 x 7 years

Finally, remember Pastor Charlie used the constant of 360 days in a year to find the mystical 14,000 days at the time of Moses and Joshua. Therefore, we replace the word "years" in the above equation with the value of 360 days per year.

69 x 7 years x 360 days/year

Hence, the years cancel out to result in a total number of days.

69 x 7 x 360 days = 173,880 days

3. The website: http://numerical.recipes/julian.html is a Julian Day and Civil Date Calculator, which gives excellent results for ancient events. If you want to verify the dates as well as the information I am writing about, you can use the calculator to do so. You can also hand calculate if you desire. This calculator will give you the Julian day number for each date. Then you can subtract the two numbers to get the value of 173,880 days. Julian day # for March 16, 445 BC = 1558962. For April 6, AD 32 = 1732842. Therefore, 1732842 – 1558962 = 173,880 days.

17. Scientific Date for Burning Down Herod's Temple

1. "Propitiousness is assigned to a propitious day and calamity to a calamitous day. As it is found and said: When the temple was destroyed, the first time, that day was immediately after the Sabbath (Sunday) . . . and so the second time (Herod's Temple also burned down on a Sunday)."

   Finegan, Jack, Handbook of Biblical Chronology, revised edition, p. 107, paragraph 203.

2. "Titus (name of the Roman General) retired into the tower of Antonia, and

resolved to storm the temple the next day, early in the morning, with his whole army, and to encamp round about the holy house; but, as for that house, God had for certain long ago doomed it to the fire; and now that fatal day was come, according to the revolution of ages; it was the tenth day of the month Lous [Ab], upon which it was formerly burnt by the king of Babylon." [Josephus, Wars of the Jews, 6.4.5].

3. Finegan, Jack, Handbook of Biblical Chronology, revised edition, Table 51, p. 108. Jack Finegan, Ph.D. in American Biblical Studies, Professor Emeritus of New Testament History and Archaeology at the Pacific School of Religion in Berkeley, California, refers to Rabbi Akiva and Rabbi Yose. The use of astronomy and the day of the week permit us to logically calculate the date of the temple's destruction by Rome occurring on Sunday, August 5, AD 70.

4. We find the idea of a New Covenant in Dan. 9:26. In English translations, this is typically "cut off" or as used in the phrase, "Messiah will be cut off and have nothing" (Dan. 9:26b). The original Hebrew word "karath" means to covenant (i.e., make an alliance or bargain, originally by cutting flesh and passing between the pieces). Thus, Jesus' death on the cross is the literal fulfillment on a future date. Although the time-based texts do not refer to the resurrection, the passages infer resurrection. Consider the question. "If Messiah must die as a sacrificial offering to prove he is the Messiah, then how will Messiah be successful?" After all, a dead Messiah will never have influence. Therefore, rising from the dead is required for the Messianic personality, who makes a blood covenant and overcomes death and hades. Refer to Strong's Concordance word no. 3772. Remember the critical question; "Should we expect to observe supernatural events (miracles) or the influence of spiritual beings on events in this universe? The logical answer is "YES" based on the principle of increasing entropy.

18. Evidence that Validates AD 32

1. Sachs, Abraham J., Steele, John M., Hunger, Hermann, Astronomical Diaries and Related Texts from Babylonia, Vol. V, Lunar and Planetary Texts, Lunar Text No. 39, pages107 to 108. Should you investigate, read instructions on page 100 on the correct interpretation of the data in regards to the length of each month.
2. The Seleucid Era for the years 88 and 89.
3. When this skewed alignment occurs, negative translations arise, which is what appears to have happened in the year AD 32. Jesus' crucifixion in the year AD 32 unveils 14,000 days in Daniel's "time-oriented" prophecy. The purpose of this prophecy was to reveal the date of the Messiah's

Notes

appearing at Jerusalem. This date on the Hebrew calendar turned out to be Palm Sunday, which was Nisan 10, the day on which the people selected a lamb for sacrificial offering (Exodus 12:3-6).

4. Beyond - The Coming Prince by Sir Robert Anderson: Table II in Appendix D reveals three or more science-based dates with unfavorable translations for the years 244 to 217 BC based on alignment to equinoxes and solstices.
5. Stern, Sacha, The Babylonian Calendar at Elephantine, p 162 - 163. Table I reveals two instances where the two calendars misalign by an entire month. The article comes from Habelt, Dr. Rudolph, Zeitschrift fur Papyrologie und Epigraphic, Bd 130 (2000), pp 159-171. I agree with both misalignments. However, this is not a scribal error. Instead, astronomical data validates a large amount of variation that was natural to the calendars before AD 360, which show the scholars use weak assumptions.
6. Beyond - The Coming Prince by Sir Robert Anderson: Table I contains this data.
7. Babylonian Chronology 626 BC to AD 75
8. Beyond - The Coming Prince by Sir Robert Anderson: False paradigm No. 3: Hebrew Calendar Month Variation
9. A year with thirteen months.
10. Search the internet for "global temperatures 2500 BC" Also of relevance was the higher temperatures before 600 BC. In that era, the beginning of Nisan 1 was much earlier in the year based on data from Babylonian Cuneiform texts. The authors of the graph comment: "Geologic evidence shows our climate has been changing over millions of years," said Harris and Mann. "The warming and cooling of global temperatures are likely the results of long-term climatic cycles, solar activity, sea-surface temperature patterns, and more." A review of the graphic depicts the influence of volcanic activity as well. The referenced graph is from the link: https://www.iceagenow.info/global-temperature-fluctuated-long-industrial-revolution-graphic/
11. Horn, Seigfried, Wood, Lynn, The Chronology of Ezra 7, 2nd Edition Revised, p 139 - 142. Originally published in the Journal of Near Eastern Studies, 1954.
12. Ibid., p 139.
13. Astronomical data from 250 to 217 BC

    Sachs, Abraham J., Steele, John M., Hunger, Hermann, Astronomical Diaries and Related Texts from Babylonia, Vol. V, Lunar and Planetary Texts, Lunar Text No. 39, pages 100 to 109. Should you investigate, read instructions on page 100 on the correct interpretation of the data in regards to the length of each month.
14. Stern, Sacha, The Babylonian Calendar at Elephantine, Table I, row C8-9, far-right column shows "1 month" discrepant. Based on Habelt, Dr.

Rudolph, Zeitschrift fur Papyrologie und Epigraphic, Bd 130 (2000), pp 159-171.
15. Horn, Seigfried, Wood, Lynn, The Chronology of Ezra 7, 2nd Edition Revised, p 142. Originally published in the Journal of Near Eastern Studies, 1954.
16. Ibid p 142
17. A reference that only uses the names of the month without a day reveals it occurred on the first day of that month (On the Stele, it is Mechir 1 and Sivan 1 when there are no day numbers).

    Good examples of this idea: Kraeling #4 shows that Tishri 25 = Epiphi 25 (actual days even though the same date). Kraeling #6 shows Pharmuthi 8 = Tammuz 8 (actual days even though the same date). Kraeling #7 shows Tishri = Epiphi (every day of the month is the same, but it is the first day of the month when written without a number). AP #20 shows Elul = Payni (same as Kraeling #7). Neh. 2:1 only mentions Nisan without the day number, which means it is Nisan 1. Support for this viewpoint comes from the travel time of Ezra. He left Persia on Nisan 1 and arrived at Jerusalem on Ab 1 (See Ezra 7:9). We find a similar period in Neh. 2:11, 6:15, with a total of fifty-five days on Elul 25. Back up fifty-five days and this supports that Nehemiah arrived at Jerusalem very close to Ab 1. Hence, the decree on Nisan 1 is credible.
18. People who agree with these two scholars claim the decree to rebuild Jerusalem happened in the year 444 BC. However, astronomical evidence shows this viewpoint is not valid.
19. Horn, Seigfried, Wood, Lynn, The Chronology of Ezra 7, 2nd Edition Revised, p 145 when AP #14 has a negative translation time of only one hour and twenty-six minutes. For that small amount of negative time, they comment that this is "unthinkable." A paradigm statement from these two scholars.
20. If you want to review all the data, please read book three in the Expect to Live Forever book series, Beyond - The Coming Prince by Sir Robert Anderson.
21. Anderson, Sir Robert, The Coming Prince, published in 1894.
22. Finegan, Jack, Handbook of Biblical Chronology, revised edition, p. 107, paragraph 203.

19. Jesus' Second Coming Promise

1. An excellent example of the unique research found in Luke's gospel is the story of Jesus as a twelve-year-old boy conversing with the learned in the temple. Where did Luke, the investigator, get this story? Maybe Luke

Notes

received it from James, the brother of Jesus. Or perhaps Luke met with Mary, the mother of Jesus, at one of the meetings in Jerusalem detailed in the book of Acts.

2. Should you want to know why I do this research and write, then please read Secrets - never heard until now - of the Book of Revelation, chapter three is about these prophetic Bible verses.

3. When was the gospel of Luke written? According to Classics scholar, Hemer, Colin J., The Book of Acts in the Setting of Hellenistic History, published 1990 by Eisenbrauns, Luke wrote the book of Acts in AD 62 in Rome. Luke wrote his gospel first, then the book of Acts. Hemer's superior knowledge of classic literature supports that Luke's gospel was written a few years before AD 62. Evidence that the Christian faith began to flourish around Rome comes from a house church covered by Mt Vesuvius eruption dated August 24, AD 79 [Sukenik, E.L., "The Earliest Records of Christianity," The American Journal of Archaeology, Vol. LI, No. 4, 1947, p27-29].

Josephus was published AD 93/94, more than thirty years after Luke wrote his gospel. Consider that Josephus was an eyewitness of the destruction of Jerusalem and the temple in AD 70. Luke's gospel records Jesus' prophecy of Jerusalem's destruction. Did Josephus read about Jesus' prophecy regarding the coming destruction of Jerusalem?

Josephus wrote about Jesus in Antiquities. Those words in Josephus correlate to Luke's gospel at the 98.8% confidence level, which strongly suggests Josephus read the gospel of Luke.

Did Josephus read and plagiarize Lukan texts regarding the road to Emmaus story?

Perhaps Josephus plagiarized Luke's gospel to include the Jewish roots of the Christian faith in Antiquities (ref: 18.3.3.63). I wrote this idea to stimulate your imagination of what may have occurred concerning Josephus. Evidence of the 14,000 days hidden in prophetic texts permits us to question scholars who place late dates on books of the Bible since anything prophetic is suspect of a late publish date.

Consider two writings that preceded Josephus and how he wrote of the destruction of Herod's Temple. First, the book of Daniel with the foretold destruction of Herod's Temple that Josephus beheld, and second, the gospel of Luke composed before AD 62 at Rome, with details that Jesus foretold of the destruction event. The high confidence level and the circumstances in Josephus' life support Josephus wrote the materials about Jesus by plagiarizing the Lukan gospel account.

4. Refer to Matt. 24:31, Rev. 11:15, Lev. 23:24-25

Notes

20. Do the Time-Based Texts Foretell Future Events?

1. A great example is the transfiguration story. Jesus takes the inner circle of three up a mountain where he transfigures in the brilliant light of the spiritual realm. The three disciples become terrified due to the brightness of seeing the glory of God. You can read this story in Matt. 17:1-8, Mark 9:2-8, Luke 9:28-36, referred to in 2nd Pet. 1:16-18, and likely in John 1:14, "We beheld his glory!"

   Some Atheist scholars teach that the belief in Jesus as the Son of God and second member of the Trinity is the result of Pagans becoming Christians in the final decades of the first century. These scholars typically do not refer to the synoptic gospels but only the gospel of John. This non-biblical viewpoint chooses to ignore the Synoptic gospel stories of the transfiguration found in all the gospels since this story portrays Jesus as being equal to an Eternal Being. Also, they ignore Old Testament texts that support the deity of the Messiah. For example, refer to Ps. 2:7, Ps. 110:1, Prov. 30:4, Dan. 7:13-14, Mic. 5:2, Isa. 9:6 and Jer. 23:5-6 as biblical texts foretelling of the person of Messiah depicted as a divine-human being. Perhaps Jeremiah's words give us the most intriguing Old Testament texts regarding the deity of the Messiah. This text uses the Tetragrammaton [a four-letter word YHVH or YHWH in Hebrew that Jewish believers will never pronounce because it is the name of G_d and is too holy for humans to pronounce.] as the name of the Messiah. Jeremiah's words strongly support the Messiah is a divine personality.

2. Jesus states, "I am the way, and the truth, and the life; no one comes to the Father, but through me" (John 14:6). Other passages support this idea. "... if you confess with your mouth Jesus as Lord, and believe in your heart that God raised Him from the dead, you shall be saved; for with the heart man believes, resulting in righteousness, and with the mouth he confesses, resulting in salvation" (Rom. 10:9-10). Muslims do not believe or accept these requirements for eternal life. The logical conclusion strongly supports that no Muslims will enter into heaven as they reject the Messiah foretold in the time-based texts.

www.ingramcontent.com/pod-product-compliance
Lightning Source LLC
LaVergne TN
LVHW041334080426
835512LV00006B/445